MOLECULAR
BIOLOGY
INTELLIGENCE
UNIT

WATER CHANNELS

Artist's schematic of CHIP28 water channel showing water pores through individual monomers in a tetrameric subunit association. See Chapter 6 for details. (Painted by Ilona H. van Hoek).

MOLECULAR
BIOLOGY
INTELLIGENCE
UNIT

WATER CHANNELS

Alan S. Verkman, M.D., Ph.D.

University of California
San Francisco

R.G. LANDES COMPANY
AUSTIN

Molecular Biology Intelligence Unit

WATER CHANNELS

R.G. LANDES COMPANY
Austin

CRC Press is the exclusive worldwide distributor of publications of the
Molecular Biology Intelligence Unit.
CRC Press, 2000 Corporate Blvd., NW, Boca Raton, FL 33431. Phone: 407/994-0555.

Submitted: August 1993
Published: October 1993

Production Manager: Terry Nelson
Copy Editor: Constance Kerkaporta

Copyright © 1993 R.G. Landes Company
All rights reserved.

Please address all inquiries to the Publisher:
R.G. Landes Company, 909 Pine Street, Georgetown, TX 78626
or
P.O. Box 4858, Austin, TX 78765
Phone: 512/ 863 7762; FAX: 512/ 863 0081

ISBN 1-57059-017-6   CATALOG # LN 9017

While the authors, editors and publisher believe that drug selection and dosage and the specifications and usage of equipment and devices, as set forth in this book, are in accord with current recommendations and practice at the time of publication, they make no warranty, expressed or implied, with respect to material described in this book. In view of the ongoing research, equipment development, changes in governmental regulations and the rapid accumulation of information relating to the biomedical sciences, the reader is urged to carefully review and evaluate the information provided herein.

Library of Congress Cataloging-in-Publication Data

Verkman, Alan S.
Water channels / Alan S. Verkman.
    p. cm. — (Molecular biology intelligence unit)
Includes bibliographical references and index.
    ISBN 1-57059-017-6
1. Water—Physiological transport. 2. Osmoregulation. I. Title. II. Series.
QH509.V48 1993
612.3'923—dc20   93-34244
                                                                                          CIP

CONTENTS

1. Overview and Historical Perspective 1

2. Water Transport Biophysics and Measurement Strategies 5

3. Cell Biology of Vasopressin-Sensitive Water Transport................................... 19

4. The Search for Water Channels.......................... 33

5. Expression and Function of the CHIP28 Water Channel 48

6. Structure and Water-Transporting Mechanism of CHIP28 60

7. Water Channel Family Relations 74

8. Tissue Distribution and Physiology of Water Channels 84

References 98

Index 114

FOREWORD

The mechanisms and physiological control of water transport across biological membranes are subjects of long-standing interest. Recent advances in the cell and molecular biology of water transport, including the identification of specific water transporting proteins, have yielded new insights into how and why water moves across cell membranes and have transformed the century-old subject of osmosis into a rapidly advancing field. The major questions to be addressed no longer involve characterization of transport phenomena, but instead focus on elucidation of the structure, biophysical mechanisms, transcriptional regulation, and intracellular processing of cloned and purified water channels.

This book was prepared over a brief period of time (July and August, 1993) to provide a state-of-the-art report on what has been learned recently about water transport and where the field is going. Although older work is cited briefly, the focus of this book is on advances made over the past five years on the biophysics, and cell and molecular biology of water transport in mammalian cell membranes. Several important related topics are not covered by this work, including fluid transporting mechanisms in epithelia and water transport in non-mammals and plants.

It is likely that advances in the understanding of water transporting mechanisms will yield new insights into membrane protein structure at the basic science level and new therapies at the clinical level. The modulation of water transport activity by pharmacological agents may provide novel treatments for hypertension, congestive heart failure and other fluid-retaining states.

I wish to acknowledge several students and fellows in my laboratory who have contributed recently to the elucidation of water transporting mechanisms — Steve Bicknese, Joachim Biwersi, Javier Farinas, Antonio Frigeri, Hajime Hasegawa, Hung Pin Kao, Tonghui Ma, Alfred N. van Hoek, Lan-bo Shi, Rubin Zhang — as well as collaborating investigators — Dennis Ausiello, Dennis Brown, Vishu Lingappa, William Skach, Jean-Marc Verbavatz and Michael Wiener. I also thank the R.G. Landes Company for their support.

—Alan S. Verkman, M.D., Ph.D.
Professor of Medicine and Physiology

CHAPTER 1

OVERVIEW AND HISTORICAL PERSPECTIVE

There has been great interest in the mechanisms by which water moves across membranes, beginning with early phenomenological studies of osmosis in porous membranes over a century ago, to the recent appreciation that specific water transporting proteins play a key role in organ physiology. The concept of selective water transporting proteins is not new. In a monograph entitled, "Osmotische Untersuchungen" (Osmotic Investigations) published in 1877,[179] Dr. W. Pfeffer writes (translated from German):

"...In animal membrane and similar membranes, capillary pores as well as narrower spaces doubtless exist. The latter probably do not allow certain salts to diosmose. Such an arrangement could in itself explain all the observed phenomena of diosmosis and swelling, but these phenomena are also consistent with the simultaneous passage of water, or of salt molecules, through membrane particles. I must leave open the question whether the latter process is involved (as seems probable)...."

There were a series of phenomenological measurements of osmotic water permeability across various animal tissues in the 1950s and 1960s. The underpinnings for much of the recent work were established in the 1950s by Solomon and collaborators,[80] who demonstrated that mammalian erythrocytes had a high-water permeability that was strongly inhibited by mercurial compounds; biophysical analysis of the water transporting pathway by measurements of temperature dependence and osmotic-to-diffusional water permeabilities suggested that the water transporting pathway consisted of an aqueous pore traversing the membrane. Soon thereafter, it was discovered that specific nephron segments in mammalian kidney also have high water permeability and may contain "facilitated" water transporting pathways. Water permeability in proximal tubule and thin descending limb of Henle were shown to be constitutively high, whereas water permeability in collecting duct is low in the unstimulated state and becomes high after exposure to antidiuretic hormone (ADH, vasopressin). Bentley, Leaf and colleagues established the amphibian urinary bladder as an easily accessible model for vasopressin-stimulated water transport.[13,138] As described in Chapter 3, a series of elegant studies led to the "membrane shuttle hypothesis" for vasopressin-regulated water permeability in which vasopressin induced the fusion of subcellular organelles containing water transporters with the apical membrane of target epithelial cells.[27,91] In response to withdrawal of vasopressin, the functional water transporting units were retrieved from the cell surface by an endocytic mechanism.

The concept of a primary active "water ATPase" was considered for many years. However in the mid-1950s, biophysical arguments based on energy expenditure ruled out the possibility of a water ATPase.[26] Water movement thus occurs passively in response to osmotic and hydrostatic driving forces produced by primary or secondary active ion and/or solute transport. It was thus predicted *a priori* that water transporters should exist on cell membranes in which water must move rapidly [relative to osmotic driving forces]. Indeed, as described in Chapter 8, the extrarenal tissue distribution of water transporting proteins, e.g., choroid plexus, ciliary body, lung alveolus, sweat gland, etc., corresponds well to predictions based on organ physiology.

The existence and molecular identity of selective water transporters, now called "water channels", has been a subject of considerable interest. By the 1970s, the idea that water transporters existed, at least in erythrocytes, had become popular. In a classic monograph entitled "Water transport in cells and tissues",[110] Dr. C.R. House writes, "... Part of the difficulty *[in explaining certain results]* is the tacit assumption that water transport occurs solely through systems of small pores about which we know very little. Even the question of their *[water transporters]* existence is not entirely free from controversy." It was unclear whether these water transporting pores consisted of protein, lipid, or some heterologous assembly of lipids, proteins and other membrane-associated components (Fig. 1.1). Early radiation inactivation and proteolysis studies raised the possibility that the water channel is small and may be composed partially of lipids. The speculation that the ubiquitous anion exchange and glucose transport proteins serve as water channels received attention. It was also proposed that many transmembrane proteins have a finite [nonselective] water permeability that might account for the high water permeabilities observed in some membranes. There have been a number of attempts to label putative water transporting proteins in erythrocytes, by radiolabeled mercurial inhibitors, and in toad urinary bladder, by chemical or antibody labeling of apical membrane and/or endosomal

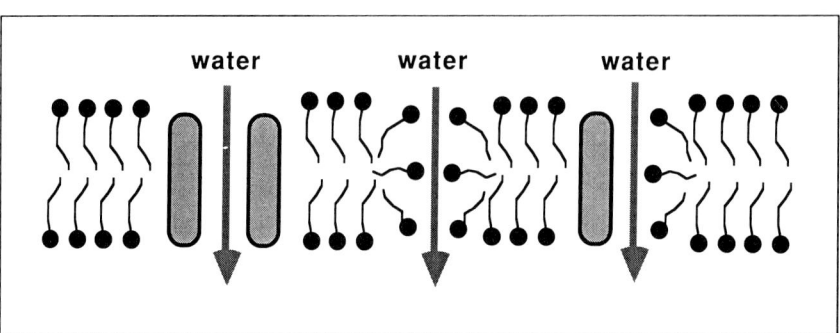

Fig. 1.1 Possible structures of biological water channels showing aqueous pores through proteins, lipids and protein-lipid assemblies.

components. The search for water channel is reviewed in Chapter 4. Although a number of candidate water transporting proteins were identified, no direct functional evidence was obtained to demonstrate that any of the candidate proteins actually transported water. The development of the oocyte expression system for water transporters and the determination of water transporter target size by radiation inactivation led to the identification of the first water transporting protein, CHIP28 (CHannel-forming Integral Protein of 28 kDa), and subsequently, the cloning of related water transporters.

The identification and characteristics of CHIP28 are described in Chapters 5 and 6. CHIP28 is a selective water transporting protein that does not pass ions and small solutes. It is very abundant in erythrocytes, kidney proximal tubule and thin descending limb of Henle, as well as a variety of extrarenal fluid-transporting tissues. There is evidence that CHIP28 is a member of a family of homologous water transporting proteins that are widely distributed in cell membranes and play an important role in cellular and organ physiology (Chapters 7 and 8).

ROLE OF WATER TRANSPORT IN KIDNEY AND AMPHIBIAN URINARY BLADDER

The need for high water permeability in certain cell membranes is well-established. Figure 1.2 shows a schematic of water permeability in various nephron segments. The high water permeability in kidney proximal tubule facilitates the near isosmotic reabsorption of glomerular filtrate driven by small osmotic gradients. The low water permeability in ascending limb of Henle and the vasopressin-regulated water permeability in kidney collecting duct are essential components of the urinary concentrating mechanism. In

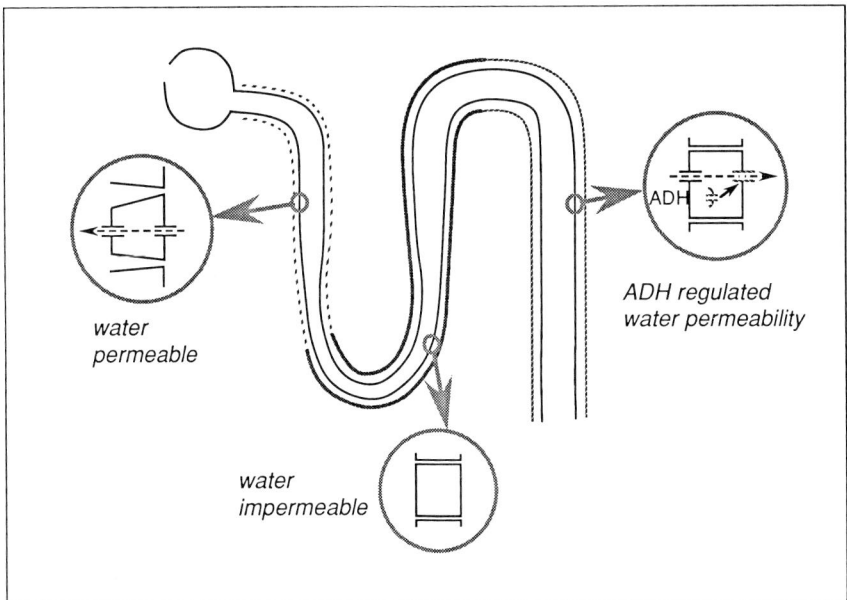

Fig. 1.2. Pathways for water transport in the kidney nephron, showing high water permeability in proximal tubule and thin descending limb of Henle, low water permeability in ascending limb of Henle, and ADH (antidiuretic hormone, vasopressin) regulated water permeability in collecting duct.

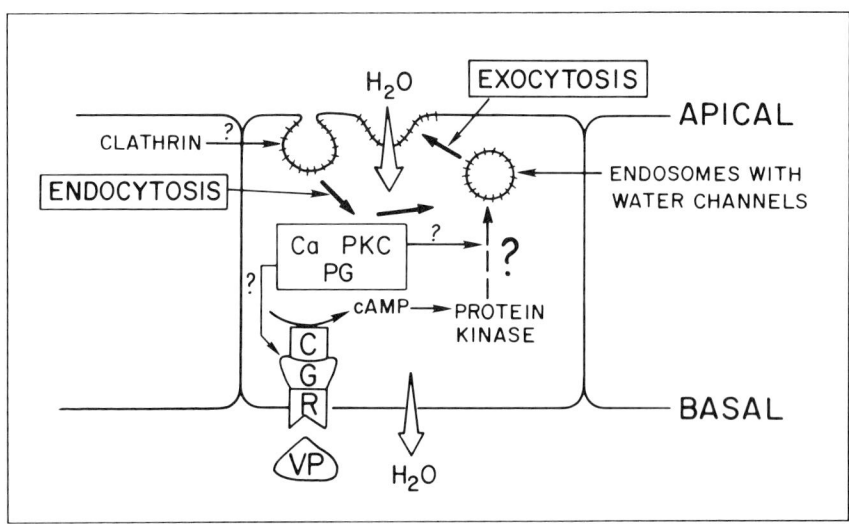

Fig. 1.3. Working hypothesis for regulation of water permeability in kidney collecting duct and amphibian urinary bladder by vasopressin. Apical membrane water permeability is regulated by the exocytic-endocytic shuttling of vesicles containing water channels between an endocytic compartment and the cell apical plasma membrane. See text for details.

the presence of vasopressin, a concentrated urine is formed by osmotically-driven movement of water from the lumen of the water-permeable collecting duct to the hypertonic medullary interstitium. In amphibian urinary bladder, vasopressin-regulated water permeability is important to control the distribution of water between the blood and urinary compartments.

In kidney collecting duct and amphibian urinary bladder, vasopressin strongly increases transepithelial water permeability. Fig. 1.3 shows a schematic of a working model for vasopressin action; additional details are provided in Chapter 3. In the absence of vasopressin, it is believed that the basolateral membrane is water permeable and the apical membrane is water impermeable, giving a low transcellular water permeability. Upon exposure of basolateral membrane (V_2) receptors to vasopressin, intracellular cAMP increases, protein kinase A is stimulated, and ultimately intracellular vesicles containing water channels fuse with the apical plasma membrane, giving high apical membrane water permeability. Upon withdrawal of vasopressin, water channels are retrieved from the apical membrane by endocytosis. There is a large body of morphological evidence correlating water permeability with the appearance of intramembrane particle aggregates (IMPs) in the apical membrane. The IMPs observed by freeze-fracture electron microscopy are believed to be the morphological equivalent of water channels.

The physiology of water transport in extrarenal tissues is reviewed in Chapter 8. It should be noted that although much of our understanding of water transporters is derived from work in erythrocytes, the significance of the high water permeability in erythrocyte membranes remains unclear. It is possible that the high water permeability is important for passage of erythrocytes through the hypertonic renal medulla and/or the narrow splenic capillaries. Erythrocyte CHIP28 might also play a role in membrane stabilization as discussed in Chapter 8.

CHAPTER 2

WATER TRANSPORT BIOPHYSICS AND MEASUREMENT STRATEGIES

The principles of water transport biophysics that are important for studies of water channel cell and molecular biology are reviewed in the first part of this chapter. For original derivations and additional details, the reader can refer to several works dedicated to nonequilibrium thermodynamics, pore theory and epithelial cell transport.[45,63,124,125] In the second part of this chapter, modern strategies to measure water permeability in membrane vesicles, reconstituted proteoliposomes, single cells, *Xenopus* oocytes and renal tubules are described and critically evaluated.

BIOPHYSICS OF WATER TRANSPORT

Table 1 lists a set of physical properties that provide a quantitative phenomenological description of water movement across a barrier separating two compartments. The barrier might be a single bilayer membrane or a complex structure consisting of multiple serial and parallel pathways for water movement. Biophysical analysis of membrane water transport provides useful information about whether functional water channels are present.

The osmotic (also called hydraulic) permeability coefficient (P_f, in units cm/s) is the most important single parameter describing water movement

Table 1. Water transporting characteristics of membranes

	Lipid pathway	Water channels
Osmotic water permeability (P_f)	<0.005 cm/s	>0.01 cm/s
Diffusional water permeability (P_d)	*	*
Ratio P_f/P_d	1	>3
Activation energy (E_a)	>10 kcal/mol	<6 kcal/mol
Mercurial inhibition	–	+ or –
Solute/proton permeability	*	*
Solute reflection coefficients	*	1 or <1

General values of P_f, P_f/P_d and E_a are given for single membranes in which water moves through a lipid pathway or through water channels.[62,246]
* indicates that no useful distinction can be made. See text for explanations.

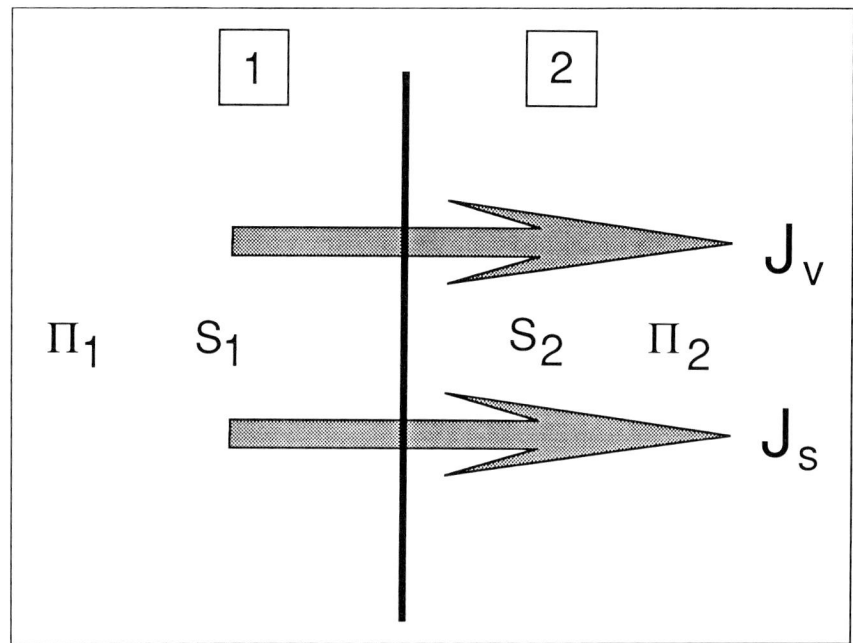

Fig. 2.1. Water and solute transport across a membrane separating two compartments. Positive volume (J_v) and solute (J_s) flux are indicated. See Equations 1 and 3 for details.

across membranes. P_f is related to the amount of volume (water) flow across the membrane in response to osmotic and pressure driving forces,

$$P_f = J_v / [S\ V_w(P_1-P_2)/RT + \Sigma\ \sigma_i\ (\Pi_{2i}-\Pi_{1i})] \qquad \text{(Eqn. 1)}$$

where J_v is volume flow (cm³/s) from compartment 1 to 2 (Fig. 2.1), S is surface area (cm²), V_w is the partial molar volume of water (mol/cm³), P is hydrostatic pressure (atm), σ_i the reflection coefficient of the ith solute, and Π_i osmolality of the ith solute (mOsm). Equation 1 indicates that hydrostatic pressure gradients ($P_1 > P_2$) and osmotic gradients ($\Pi_2 > \Pi_1$) drive volume from compartments 1 to 2. Each solute contributes to the effective osmotic gradient in proportion to its reflection coefficient σ_i. The reflection coefficient is a phenomenological parameter derived from nonequilibrium thermodynamics. Mechanistically, σ_i can be defined as,

$$\sigma_i = 1 - (P_s/P_f)(V_s/V_w) - f \qquad \text{(Eqn. 2)}$$

where P_s is the solute permeability coefficient (cm/s), V_s is the partial molar volume of solute (mol/cm³) and f is a frictional term that describes interactions between water and solutes that occur only if they traverse the same pathway through the membrane; f is zero if water and solute move through separate pathways. By the Onsager reciprocity relation, σ_i describes both the efficacy of solute i to induce osmotic water movement across a membrane (Eqn. 1), and the "solvent drag" of solute i induced by volume movement. For an impermeant solute σ_i approaches unity, whereas σ_i approaches zero and can be negative when P_s is very high. Equations 1, together with 3 (below), are known as the Kedem-Katchalsky nonequilibrium thermodynamic equations for coupled water and solute movement,[126,127]

$$J_s = P_s ([S_1]-[S_2]) + J_v (1-\sigma_i) <S> \qquad \text{(Eqn. 3)}$$

where J_s is solute flux (mol/cm²s) from compartments 1 to 2, [S] is solute concentration, and <S> is the effective solute concentration in the membrane; <S> is often assumed to equal the mean value $([S_1]+[S_2])/2$. The first term in Equation 3 describes concentration-driven solute movement and the second term describes the solvent drag of solute. There are separate equations for solute flux corresponding to each solute; the equations must incorporate electroosmotic phenomena and electrochemical driving potentials when charged solutes are considered.[6]

For a simple oil membrane in which water permeability occurs by a nonfacilitated solubility-diffusion mechanism, P_f is equal to $D_w K_w V_w/dV_{oil}$,[63] where D_w is the diffusion coefficient of water in the oil phase (cm²/s), K_w is the partition coefficient of water between the aqueous and oil phases, d is the thickness of the oil membrane and V_{oil} is the molar volume of the oil phase. For a membrane containing macroscopic (wide) pores allowing fully developed Poiseuille flow (as occurs in water pipes),

$$P_f = (nRT\ \pi r^4)/(8ALV_w\eta) \qquad \text{(Eqn. 4)}$$

where n is the number of pores, r is pore radius, A is membrane area, L is pore length and η is viscosity in the pore. Note the fourth power dependence of P_f on pore radius. Equation 4 is probably accurate for wide pores of >10 nm radius. The P_f, A and n terms are often combined to give the single channel water permeability, p_f (cm³/s), defined by $p_f = P_f A/n$. The single channel water permeability provides a quantitative description of volume flux through individual water pores or channels.

For single-file transport of water through narrow pores, a very different equation relates p_f to the pore properties. Assuming that water density in the pore is the same as that in bulk solution,

$$p_f = \pi r^2 D_w^{(o)}/L \qquad \text{(Eqn. 5)}$$

where $D_w^{(o)}$ is the diffusion coefficient of a water molecule if it were the only molecule in the pore. Equation 5 probably applies to narrow pores of radius <0.5 nm where water must move single file. Note that flow equations have not been developed for intermediate radii; in addition, there is no reason to believe that biological water channels should conform to the idealized assumption of uniform right cylindrical pores or channels.

For a simple membrane in which membrane properties do not change with osmolality and water flow, water transport is symmetric and P_f is independent of osmolality and osmotic gradient size.[162,220] In more complicated barriers, it is possible to have asymmetric water flow and osmotic gradient-dependent P_f values.[6,16,198,298] For example, if water flow in one direction changes the shape of the lateral intercellular space in an epithelium, then P_f would depend on the direction of the osmotic gradient. For two membranes (or water transporting pathways) in parallel, P_f values are additive ($P_f = P_{f1} + P_{f2}$); for membranes in series, P_f values add reciprocally ($1/P_f = 1/P_{f1} + 1/P_{f2}$). For biological membranes, P_f of < 0.005 cm/s generally indicates that water moves by a solubility-diffusion mechanism without water channels,[61,62] whereas P_f >0.01 cm/s suggests the presence of water channels.

A second parameter describing water movement is the diffusional water permeability coefficient P_d (cm/s). Diffusional water permeability (like diffusional solute permeability) is described by the Fick equation as the movement of tracer-labeled water from compartments 1 to 2 in the absence of an osmotic or pressure gradient,

$$P_d = J_d/\{S\,([^*H_2O]_1 - [^*H_2O]_2)\} \qquad \text{(Eqn. 6)}$$

where J_d is the flux of labeled water and $[^*H_2O]$ is the concentration of labeled water. P_f and P_d are fundamentally different parameters that require very different measurement approaches (see below). P_f and P_d are equal for a simple membrane that does not contain water channels and in the absence of unstirred layers. The ratio P_f/P_d can be greater than unity when water moves through a wide pore or narrow channel,[142,162,264] or when P_d is artifactually low because of unstirred layers. For a wide right cylindrical pore, $P_f/P_d \sim 1 + RTr^2/8\eta D_w V_w$; for a narrow channel in which single-file movement of water occurs, $P_f/P_d \sim N$, the number of water molecules in the channel.[63] It is stressed that this formulation requires a number of assumptions and applies to idealized geometries that probably do not exist in proteinaceous water channels.

The activation energy (E_a, kcal/mol) for water permeability is a useful parameter for identifying water channels in membranes. E_a is determined by the Arrhenius equation from the temperature dependence of P_f,

$$\ln P_f = -E_a/RT + C \qquad \text{(Eqn. 7)}$$

where C is an integration constant related to activation entropy. E_a is generally >10 kcal/mol for water movement by a channel-independent solubility-diffusion mechanism and <6 kcal/mol for water movement through aqueous pores. The high E_a for water movement through lipid is probably related to the formation and breaking of hydrogen bonds as water moves from the aqueous to membrane phases and back to the aqueous phase. Movement of water through aqueous pores should be associated with fewer bond formation and breaking events. Several caveats should be noted in the interpretation of activation energies. In complex membranes consisting of multiple serial and/or parallel barriers to water transport, the Arrhenius ln P_f vs. 1/T relation can be nonlinear. For example, in membranes containing parallel channel- and lipid-mediated pathways for water movement, E_a may be low at low temperatures, where water moves primarily through channels, and high at high temperature, where water moves primarily through lipid. If water movement is rate-limited by unstirred layers and not by the membrane, measured E_a can equal the activation energy for solute diffusion in aqueous solutions (~5 kcal/mol), even if water moves through membrane lipid. Further, E_a might be high (>6 kcal/mol) when water moves through narrow or gated channels in which significant interactions between water molecules and the channel wall can occur.

The ability of mercurial sulfhydryl compounds to inhibit water permeability in certain biological membranes, including erythrocytes and kidney tubules, has been taken as evidence for proteinaceous water channels. As described in Chapter 5, the specific site of action of mercurial inhibitors is now established for certain water channels. Mercurials do not affect water movement through the lipid portion of the membrane bilayer. As described in Chapter 4, other proteins (e.g., glucose transporter, CFTR) can pass some water but are not inhibited by mercurials. In addition, there may be other selective water transporting proteins that are not mercurial-sensitive.

Several related transporting properties of membranes are useful in the evaluation of water channels. For wide water pores which can pass solutes, the dependence of solute permeability on solute size can give useful information about apparent pore size.[80,183] If solute reflection coefficients are low, then equation 2 indicates the existence of a common aqueous pathway for water and solutes, implying the presence of water channels. The artificial pore-forming agents amphotericin B and gramicidin A transport water as well as cations and protons.[63,109] Although water channels identified in biological membranes appear to be selective for transport of water (Chapter 5), there may be other water transporting proteins with less selectivity.

MEASUREMENT OF WATER TRANSPORT

Light Scattering Measurement of Osmotic Water Permeability in Cells and Vesicles

The light scattering method provides a rapid and direct approach to measure osmotic water permeability in osmotically-responsive, sealed membrane vesicles. Light scattering was developed initially for water permeability measurements in erythrocytes,[147,162] and subsequently applied in membrane vesicles,[233,253,281,282] proteoliposomes[237,297] and cell suspensions.[39,147,159] The method is based on the dependence of scattered light intensity on vesicle volume. Experimentally, a vesicle suspension is mixed rapidly with an anisosmotic solution (generally hypertonic) to generate a transmembrane osmotic gradient. Mixing is usually accomplished by a stopped-flow apparatus in which small volumes (0.02-0.1 ml) of two solutions are mixed together in <1 ms. As osmotic water flux occurs, the time course of scattered light intensity provides an instantaneous measure of vesicle volume. For 90° light scattering measurements performed on small membrane vesicles with diameters of 0.1-1 micron, scattered light intensity generally decreases with increasing vesicle size. The theory predicting the intensity of scattered light from vesicles incorporates angle- and wavelength-dependent interference patterns, vesicle geometry, and membrane and solution refractive index. The relatively complex theory is neither required nor utilized in experimental measurements of water permeability.

Figure 2.2 (left) shows a stopped-flow light scattering experiment in which human erythrocytes were subjected to an inwardly-directed gradient of NaCl. The time course of increasing scattered light intensity represents osmotic water efflux and cell shrinkage; osmotic water permeability was inhibited strongly by pCMBS as shown by the slower time course. To calculate P_f, knowledge is required of the light scattering time course, vesicle surface-to-volume ratio, and the dependence of scattered light intensity on

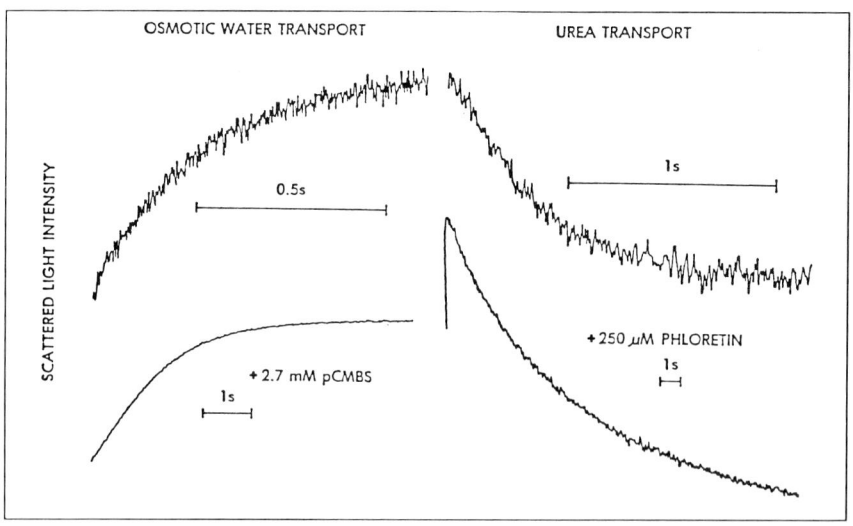

Fig. 2.2. Stopped-flow light scattering measurement of osmotic water and urea permeability in human erythrocytes. Erythrocytes were subjected to 500 mOsm inwardly-directed gradients of NaCl (left) or urea (right). Where indicated, the transport inhibitors pCMBS or phloretin were added 5-10 min prior to measurement.

vesicle volume. The differential equation describing the time course of scattered light intensity in response to an osmotic gradient of a membrane-impermeant solute can be derived from equation 1, the equation for conservation of intravesicular solute [$C_i(t)V(t) = C_i(0)V(0)$, where $C_i(t)$ is solute concentration and $V(t)$ volume], and an assumed linear relationship between scattered light intensity $I(t)$ and volume [$V(t)/V(0) = AI(t) + B$, where A and B are constants],[253]

$$dI(t)/dt = P_f V_w [S/V(0)]/A \; [C_i(0) \cdot (AI(t)+B)^{-1} - C_o] \quad \text{(Eqn. 8)}$$

where $S/V(0)$ is surface-to-volume ratio (cm^{-1}) and C_o is external osmolality. Equation 8 is subject to the initial condition $I(0) = (1-B)/A$. $dI(t)/dt$ is measured, C_o and $C_i(0)$ are known, and P_f, A and B are fitted. $S/V(0)$ is generally estimated by negative staining electron microscopy or quasi-elastic light scattering. P_f is then calculated by a 3-parameter nonlinear regression[253] or estimated by approximate formulae that assume exponential relaxation.[233] The assumption that scattered light intensity is linearly related to vesicle volume has been shown to be reasonably valid for a variety of vesicle and cell systems.[160,253,256] If necessary, the scattering vs. volume relation can be determined empirically (and incorporated into equation 8) by performing light scattering measurements at a series of different osmotic gradients.[114] To measure the activation energy E_a of osmotic water permeability, P_f is determined at a series of temperatures to construct an Arrhenius plot of ln P_f vs. $1/T$ as described by equation 7.

The light scattering method also provides a rapid approach to measure solute permeabilities. Vesicles are subjected to an inwardly-directed solute gradient and the biphasic time course of scattered light intensity is analyzed by the Kedem-Katchalsky equations. Figure 2.2 (right) shows the effect of a urea gradient on light scattering in human erythrocytes. There is rapid osmotic water efflux, vesicle shrinkage and increasing scattered light intensity, followed by a slower phase of decreasing scattered light intensity corresponding to urea and water influx; urea permeability was strongly inhibited by phloretin. In early studies, attempts were made to analyze the detailed shape of the biphasic light scattering curve to deduce solute reflection coefficients; however, uncertainties in time-dependent changes in refractive index and other assumptions in the light scattering method generally make the analysis impossible.[162] A "null-point method" which does not depend on refractive index has been used in several studies.[174,256] Vesicles containing a membrane-impermeable solute (unity reflection coefficient) are mixed with solutions containing various concentrations of a "test" solute with unknown reflection coefficient. The concentration of test solute giving zero initial osmotic water flow (flat initial light scattering vs. time curve) is determined. Solute reflection coefficient is then equal to the quotient of the concentrations of impermeable and test solutes. An alternative method to determine solute reflection coefficient based on solvent drag[174] is described in Chapter 4.

An advantage of the stopped-flow light scattering technique is that very rapid rates of osmotically-induced water transport can be measured easily using relatively small samples. However, because all osmotically active vesicles in the sample contribute to the light scattering time course, the sample must be relatively pure. In addition, the dependence of scattered light intensity on refractive index and vesicle shape sometimes confounds the quantitative determination of P_f. To overcome these difficulties, a stopped-flow fluorescence quenching method was developed as described below.

Measurement of Osmotic Water Permeability by Fluorescence Quenching

A fluorescence quenching assay permits the measurement of water permeability in selective vesicle populations, such as endosomes, that are fluorescently labeled by fluid-phase indicators.[38] It is well established that certain fluorophores with overlapping fluorescence excitation and emission spectra undergo "fluorescence self-quenching"; as fluorophore concentration increases, fluorescence intensity decreases by a nonradiative energy transfer mechanism. Experimentally, vesicles loaded with a high concentration of fluorophore are subjected to an inwardly-directed osmotic gradient in a stopped-flow apparatus (Fig. 2.3, top). As vesicle volume decreases due to osmotic water efflux, the increased intravesicular fluorophore concentration causes self-quenching and decreased fluorescence. The transport of nonelectrolytes such as urea can be measured from the time course of fluorescence in response to a urea gradient (Fig. 2.3, bottom)

Figure 2.4 shows the time course of decreasing fluorescence in renal apical vesicles loaded with 6-carboxyfluorescein. The fluorescence signal increases as the size of the inwardly-directed osmotic gradient increases. In analogy to the approach used to determine P_f in light scattering measurements (see above), P_f in fluorescence quenching measurements is determined from the time course of fluorescence, vesicle surface-to-volume ratio, and the relationship between fluorescence and vesicle volume. The fluorescence vs. vesicle volume relationship is determined from self-quenching measurements performed for a series of different osmotic gradients. Similarly, because vesicle fluorescence provides a quantitative and instantaneous

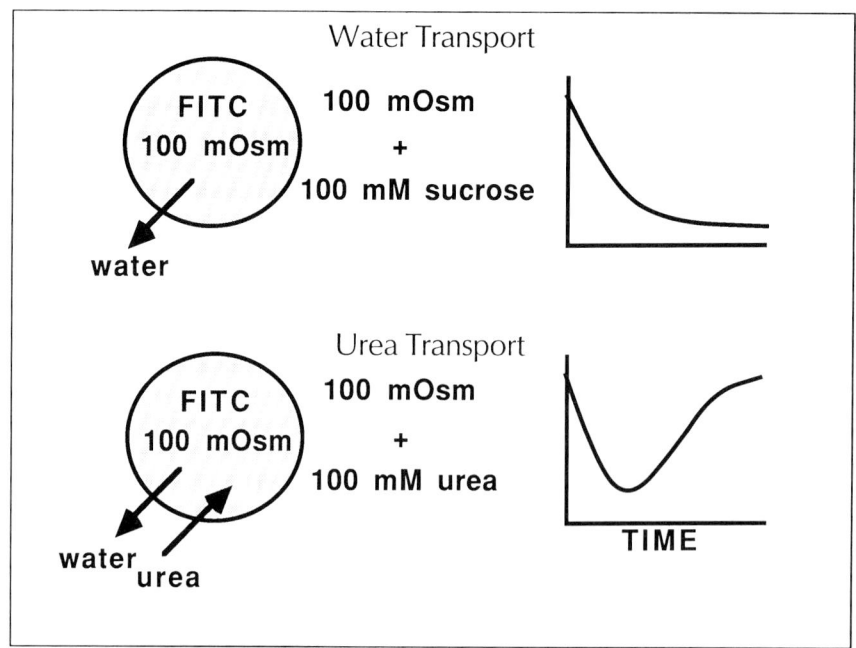

Fig. 2.3. Measurement of osmotic water and urea transport by stopped-flow fluorescence quenching. Vesicles or endosomes are loaded with a fluorophore that undergoes fluorescence self-quenching (e.g., FITC) and mixed rapidly with hyperosmotic solutions. An impermeant solute (e.g., sucrose) is used to measure water transport. The time course of fluorescence provides an instantaneous quantitative measure of vesicle volume. See text for further explanation.

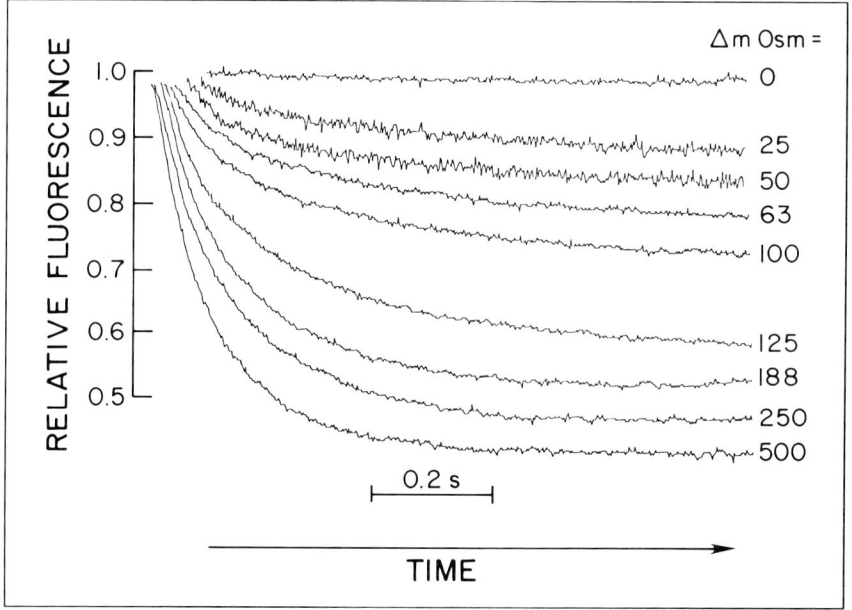

Fig. 2.4. Measurement of osmotic water permeability in apical membrane brush border vesicles from rabbit kidney by a stopped-flow fluorescence quenching method. Vesicles were labeled with carboxyfluorescein and subjected to osmotic gradients from 0 to 500 mOsm in a stopped-flow apparatus. Osmotic water efflux causes vesicle shrinkage, increased intravesicular fluorophore concentration, and instantaneous fluorescence self-quenching. (From ref. 38.).

measure of vesicle volume, solute permeability and reflection coefficients can be deduced from analysis of biphasic fluorescence curves and null-point determination, respectively. There was quantitative agreement between permeabilities obtained by the light scattering and fluorescence quenching methods when applied to the same system.[38]

A unique advantage of the fluorescence quenching method is the ability to measure osmotic water permeability in a small subpopulation of fluorescently labeled vesicles contained in a mixed population of nonfluorescent vesicles. For vesicles derived from biological membranes, fluorescence self-quenching occurs at >50-fold lower intravesicular concentrations of fluorophore (<0.1 mM fluorescein) than is predicted from self-quenching studies performed in aqueous solution. As described in Chapters 3 and 4, fluorescence quenching was used to measure water and solute permeabilities in endosomes containing water channels from proximal tubule,[283] kidney collecting duct,[139,257] toad urinary bladder[95,204] and a variety of cultured cells.[297] In addition, information about the colocalization of water channels and proton pumps in endosomes was obtained by utilizing simultaneously the dependences of carboxyfluorescein fluorescence on volume and pH.[271,283]

Approaches to Measure Diffusional Water Permeability

Diffusional water permeability is the diffusional movement of "labeled" water in the absence of an osmotic gradient. Because diffusional water permeability is not associated with net volume movement, water molecules must be "tagged" and followed as they move between two or more compartments. Several methods have been proposed and developed to tag water. Older measurements of P_d utilized tritiated water, ^3H_2O;[23,151] however, very rapid mixing and separation devices are required because of the short equilibration times associated with diffusional water movement, e.g., ~10 ms diffusional exchange time for water in erythrocytes. P_d can be measured with D_2O, by making use of isotopic differences in fluorophore quantum yields[134] or infrared absorbance. Alternatively, P_d can be measured by nuclear magnetic resonance (NMR) methods which tag H_2O molecules magnetically. The details and limitations of the several useful methods to measure P_d are described below.

NMR has been used to measure erythrocyte water permeability for decades.[3,10] The most widely utilized approach is to add a paramagnetic compound such as Mn to a cell suspension to decrease proton T_1 and T_2 relaxation times in the extracellular compartment. In the absence of diffusional water exchange, two relaxation times are measured: a slow relaxation (e.g., 1 s) for protons in water molecules in the intracellular compartment, and a fast relaxation (e.g., 2 ms) for water molecules in the extracellular compartment containing the paramagnetic compound. The pre-exponential amplitudes for the biphasic decay of magnetization depend upon the relative compartment volumes. As P_d increases, there is coupling of the compartments resulting in a decrease in the slow relaxation time. When magnetically labeled intracellular water molecules diffuse into the extracellular space, they demagnetize rapidly. The mathematical formalism relating magnetization decay/relaxation times to P_f involves eigenvalue and eigenvector determination.[280] An application of NMR to determine basolateral membrane P_d in suspended kidney proximal tubules is given in Chapter 4.

The NMR method can be utilized to measure diffusional exchange times that are slower than the relaxation time of the extracellular buffer in the presence of paramagnetic quenchers. With the use of potent quenchers such as Gd and ferrite, this time can be <1 ms. The use of chemical shift reagents may further reduce this time. However, a major disadvantage of the NMR method is the requirement of very large samples (generally >0.5 ml of >20% packed cells). Although sample size is not a problem for erythrocytes, it is a major problem for membrane and cell preparations; in addition, the high cell concentrations required and the added paramagnetic quenchers cause cell toxicity and membrane leakiness.

To overcome some of these difficulties, a fluorescence approach was developed for measurement of water diffusion based on the isotopic sensitivity of fluorophore quantum yield.[134] After screening a series of compounds, it was found that the quantum yield of aminonapthalene trisulfonic acid (ANTS) increased by >three-fold when H_2O was replaced by D_2O. ANTS is membrane-impermeable and has good optical properties including high molar absorbance and quantum yield (in D_2O), and a large Stoke's shift. Experimentally, P_d is measured in ANTS-labeled vesicles or cells by stopped-flow fluorimetry in which cells in a buffer containing H_2O are mixed with an isosmotic buffer containing D_2O. As H_2O-D_2O exchange occurs, ANTS fluorescence increases. Fluorescence measurements of P_d have been carried out in liposomes and erythrocyte vesicles; in addition, P_f and P_d have been determined at the same time by simultaneous measurements of scattered light intensity and ANTS fluorescence.[284] Light scattering has also been used to measure P_d based on the difference in refractive indices of H_2O vs. D_2O, however the signals are very small and subjected to confounding factors that influence refractive index.[137]

Measurement of Water Permeability in Single Cells

Measurement of water permeability in single adherent cells in culture (or after immobilization on a coverglass by polylysine) may be required when there is cell heterogeneity because of the presence of multiple cell types in a primary culture, or variability in expression of transfected cDNAs. To measure osmotic water permeability, the time course of cell volume is measured in response to a rapid change in bathing solution osmolality. Rapid exchange (<50 ms) of perfusion solutions can be accomplished by a laminar flow channel design.[59] Several methods have been used to follow cell volume. For some cells with favorable geometry, e.g., well-separated dome shape, light scattering provides a signal that changes with cell volume.[59,64] However, the signals are generally small and sensitive to nonvolume factors including solution refractive index. As described in Chapter 4, light scattering has been used to measure P_f in cultured J774 macrophages. A simple and potentially more general approach was reported recently by Muallem et al.[164] Cell cytoplasm was labeled fluorescently, and the fluorescence of a small area of the cell was monitored. The total number of fluorophores in the measuring area is sensitive to cell volume if changes in cell shape accompany changes in cell volume. The measurement of fluorescence in a thin (~0.8 micron) cell slice by confocal microscopy may improve detection of volume changes by this approach.

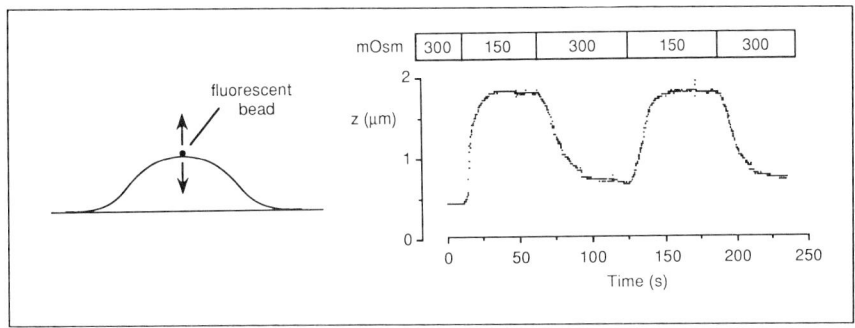

Fig. 2.5. Measurement of osmotic water permeability in adherent A431 cells by tracking the position of a fluorescent bead. Left: Schematic of cell with bead attached to the external surface. Right: Time course of bead movement in the vertical z-direction in response to changes in bath osmolality.

A novel method to measure single cell P_f was developed recently based on single particle tracking.[123] Fluorescently-labeled polystyrene beads are immobilized on the cell surface and the vertical position of a single bead is followed in response to changes in perfusate osmolality. Cell images are recorded in 50-100 ms by a cooled CCD camera and bead position is determined to ~25 nm accuracy by analysis of images using an asymmetric point-spread-function. Figure 2.5 shows the rising and falling of a rhodamine-labeled polystyrine bead, corresponding to cell swelling and shrinking, on an A431 cell in response to osmotic gradients. The method is quite general and can be used to measure water permeability in several cells at the same time.

Continuous Measurement of Water Permeability in Kidney Tubules

The conventional approach to measure osmotic water permeability in isolated perfused kidney tubules is the timed collection of fluid perfused through the tubule lumen.[2] An impermeable tracer (e.g., ^{3}H-inulin) is perfused through the lumen and a transepithelial osmotic gradient is established by bathing the tubule in hypoosmotic or hyperosmotic solutions. A schematic of the measurement approach is given in Figure 2.6. For a hyperosmotic bathing solution, solute-free water is extracted along the length of the tubule and the tracer concentration increases. P_f is calculated from the tracer concentration, lumen flow, lumen and bath osmolalities, and tubule length and surface area.[2] The measurement of P_f in isolated tubules is technically challenging because both the proximal and distal ends of the tubule must be cannulated for fluid collections; the time between data points is generally minutes because of the time required to change collection pipettes.

A simple fluorescence method was developed to measure P_f continuously in perfused tubules.[132] Rather than collecting radioactively labeled fluid, the tubule lumen was perfused with a membrane-impermeable fluorescent indicator. The change in indicator concentration was determined from the luminal fluorescence at the distal segment of tubule. Figure 2.7 shows the distal side of a perfused collecting duct from rabbit kidney; the fluorescence from a small spot near the holding pipette was recorded. Fluorescence at a single wavelength can be measured for a dilute fluorophore,[133,135] or at two wavelengths (ratio imaging) when two fluorophores are used, one of which undergoes concentration-dependent self-quenching.[131] As described in Chapter 3, this approach was used to determine the kinetics of vasopressin-induced turn-on and turn-off of P_f in kidney collecting duct,[135] and the influence of transcellular volume flow on rates of endosomal trafficking.[133]

A similar strategy was used to measure P_d continuously in perfused tubules.[134] The conventional method to determine P_d is to measure the loss of ^3H_2O from the luminal perfusate to the collected fluid. P_d is determined

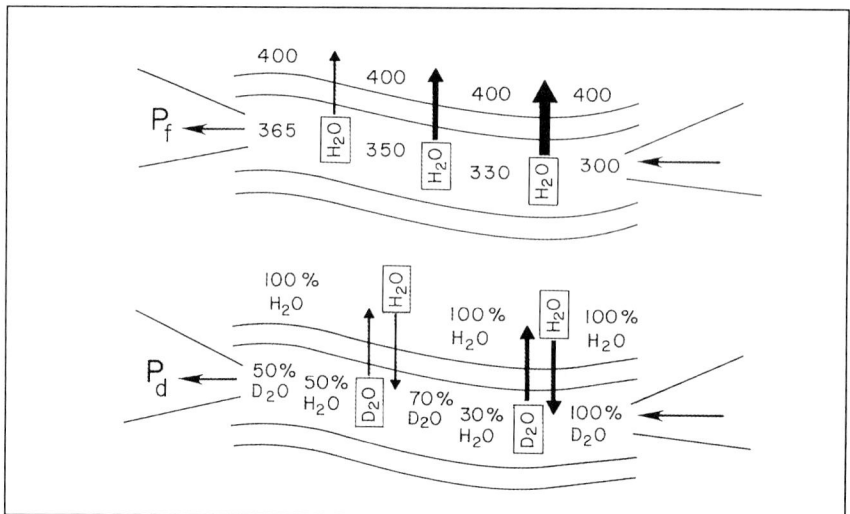

Fig. 2.6. Measurement of osmotic and diffusional water permeability in perfused kidney tubules. The lumen is perfused at constant flow (typically 5-50 nl/min) from right to left. For measurement of P_f, the osmolality (or equivalently, the concentration of a membrane-impermeant tracer) at the distal end of the tubule is measured when a transepithelial osmotic gradient is imposed. For measurement of P_d, water exchange is measured using ^3H_2O or D_2O. See text for details.

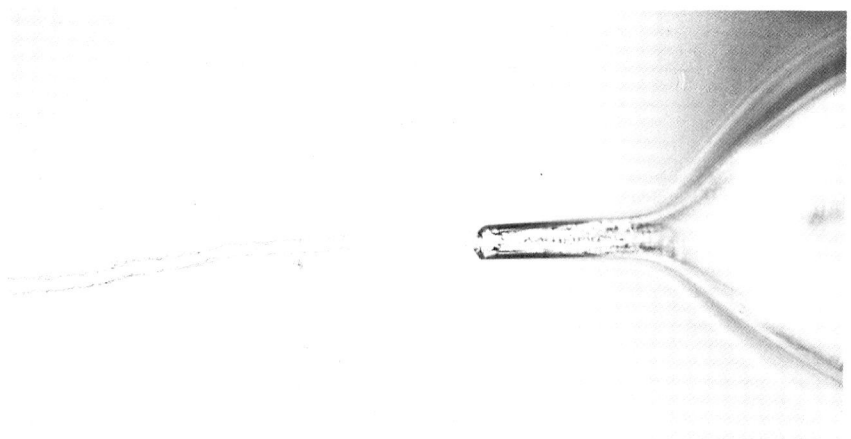

Fig. 2.7. Measurement of water permeability in an isolated perfused cortical collecting duct from rabbit kidney. The tubule lumen was perfused right-to-left with the impermeant dye fluorescein sulfonate. A small spot near the holding pipette at the distal end of the tubule is illuminated and fluorescence is recorded continuously. See text for details.

by ^3H_2O concentrations, lumen flow, and tubule length and surface area. P_d was measured continuously by replacing luminal ^3H_2O by the impermeable fluorophore ANTS, and lumen or bath H_2O by D_2O. The fluorescence of ANTS at the distal end of the tubule provided a quantitative measure of H_2O/D_2O content and thus of P_d. The fluorescence method was used to measure the kinetics of P_d in kidney collecting duct in response to vasopressin stimulation.[134]

Water Transport in Xenopus Oocytes

The *Xenopus* oocyte has been utilized extensively as an expression system for heterologous mRNA and cRNA transcribed from cloned cDNA.[207] Oocytes are microinjected with mRNA or cRNA (or sometimes injected intranuclearly with cDNA) and incubated at 18°C for one to five days for protein expression. Translated membrane proteins are generally modified and targeted to the oocyte plasma membrane so that functional measurements of transporter expression and properties are possible. Although there have been many studies of oocyte swelling over the last four decades, the first measurement of water permeability of oocytes expressing proteins was reported by Fischbarg and colleagues for the glucose transporter, GLUT1.[65] Subsequently the oocyte has been used to express mRNA encoding water channels,[58,222,294] cRNA encoding cloned water channels,[69,73,104,185,295] and other membrane proteins with pore properties[65,105,292] as described in Chapters 4, 5 and 7.

The original swelling assay involved enzymatic defolliculation of oocytes with collagenase, dilution of the extracellular solution with distilled water, and estimation of oocyte volume by measurement of two orthogonal oocyte diameters on a video monitor every one to five minutes.[65] To measure the oocyte swelling at early times with one second time resolution, an quantitative imaging approach was subsequently developed.[294,298] The oocyte was mounted in a temperature-jacketed chamber on the stage of an inverted epifluorescence microscope and illuminated from above with dim monochromatic light. The shadow cast by the oocyte (Fig. 2.8, left) was recorded by a charged coupled device (CCD) camera and digitized. A computational "mask" was then applied numerically to assign a value of 1 to all image pixels within the boundary of the oocyte and a value of 0 to other pixels (Fig. 2.8, right). A pixel sum provides a quantitative measure of relative oocyte cross-sectional area, which was con-

Fig. 2.8. Visualization of a Xenopus oocyte by a charged coupled device camera. Left: The oocyte casts a dark shadow by transmission light microscopy. Right: The image is numerically masked so that pixels within the oocyte are white and pixels outside the oocyte are black. Scale bar is 0.5 mm. (From ref. 298.)

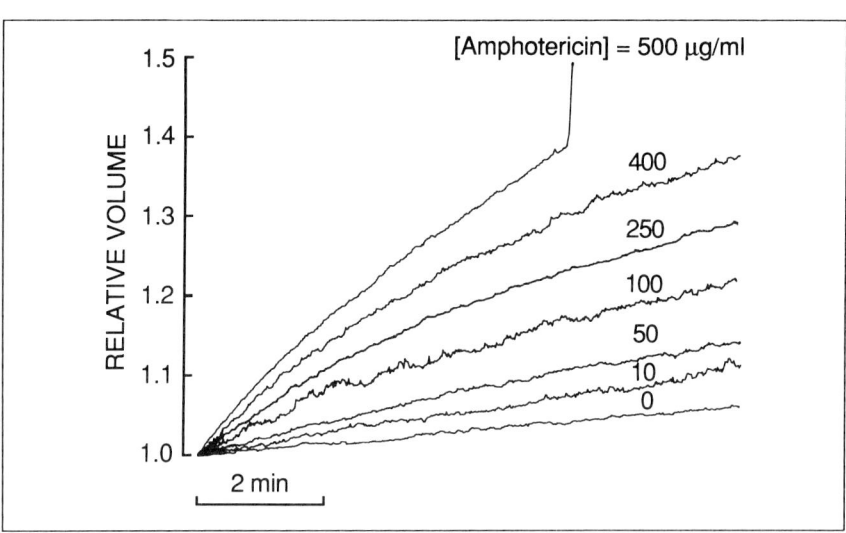

Fig. 2.9. Time course of oocyte swelling in response to a 20-fold dilution of the extracellular buffer with distilled water. Relative volume was measured in one second intervals by a quantitative imaging method as described in the text. Oocytes were incubated with the indicated concentrations of the pore-forming agent amphotericin B prior to and during the measurement. The oocyte burst at ~5 min for the highest concentration of amphotericin B. (From ref. 298.)

verted to oocyte volume by assuming proportionality between volume and (area)$^{3/2}$. Figure 2.9 shows data for the time course of oocyte swelling in response to a 20-fold dilution of extracellular medium (Barth's buffer, originally ~200 mOsm). Oocytes swelled linearly at early times; the rate of swelling increased strongly upon addition of the pore-forming agent amphotericin B. Oocyte P_f was calculated from the initial rate of swelling, $d(V/V_o)/dt$, by the relation $P_f = [d(V/V_o)/dt]/[(S/V_o)V_w(Osm_{out}-Osm_{in})]$, where $S/V_o = 50$ cm^{-1}, $V_w = 18$ cm^3/mol, $Osm_{in} = 200$ mOsm, and Osm_{out} is determined from the dilution factor. The imaging method could be used to measure increases in oocyte volume of < 1% with one second time resolution.

P_f in defolliculated oocytes measured at 25°C was quite low at 8 x 10^{-4} cm/s and E_a was high at 10.2 kcal/mol.[298] P_f was independent of osmotic gradient size and not sensitive to mercurials. The low endogenous water permeability and the absence of water channels make the oocyte an ideal expression system for water channels. It should be pointed out that swelling measurements of osmotic water permeability in defolliculated oocytes are required to quantify water permeability. Native oocytes do not swell or shrink easily because of the follicular cell layer, defolliculated oocytes do not shrink easily, and oocytes with the vittaline membrane removed do not remain spherical. P_d measurements do not give information about water permeability in the oocyte plasma membrane because diffusional transport of water is unstirred layer limited. ^3H_2O uptake measurements gave a P_d of 3 x 10^{-4} cm/s at 25°C with E_a of 6.5 kcal/mol. Whereas P_f increased by >10-fold with 0.5 mg/ml amphotericin B, P_d increased by <2-fold.[298] In addition, because of the high E_a for endogenous water permeability and the low E_a expected for water channels, the functional expression assay is most sensitive at low temperature. Oocyte swelling assays are therefore performed routinely in our laboratory at 10°C.

CHAPTER 3

CELL BIOLOGY OF VASOPRESSIN-SENSITIVE WATER TRANSPORT

Vasopressin-stimulated water permeability in mammalian kidney collecting duct and amphibian urinary bladder is an essential component of osmoregulatory mechanisms. For nearly 40 years it was known that water permeability in certain epithelia is increased by the antidiuretic hormone vasopressin. Several recent reviews have summarized the important contributions that led to the "membrane shuttle" hypothesis for regulation of water permeability by vasopressin.[27,28,91,246,251,255,265,289] In the 1960s, isolated perfused tubule techniques were used to characterize vasopressin-stimulated water permeability in collecting duct.[84] In the 1970s, the first freeze-fracture electron micrographs showed intramembrane particle aggregates whose appearance and location correlated with vasopressin stimulation.[17,22,29,92] An important role of the cell cytoskeleton in the vasopressin hydroosmotic response was implicated.[108,111] In the late 1970s and early 1980s, additional morphological studies indicated that vasopressin stimulation of water permeability paralleled the fusion of intracellular vesicles, containing intramembrane particles, with the cell apical plasma membrane.[165,268] More recent studies have focused on the cell biology of membrane trafficking in vasopressin-sensitive epithelia, functional measurements on subcellular fractions, and the identification of putative components involved in the vasopressin hydroosmotic response. The biochemical and molecular approaches to identify putative vasopressin-sensitive water channels are reviewed in Chapters 4 and 7; this chapter is focused on recent functional studies of water transporting mechanisms in vasopressin-responsive epithelial cells.

VASOPRESSIN REGULATION OF TRANSEPITHELIAL WATER PERMEABILITY AND THE MEMBRANE SHUTTLE HYPOTHESIS

The kinetics of vasopressin action on water permeability in rabbit cortical collecting duct is shown in Figure 3.1.[135] Transepithelial osmotic water permeability P_f was measured continuously by a microfluorimetric method as described in Chapter 2. Prior to addition of vasopressin, P_f was stable and low (~10 x 10^{-4} cm/s). In response to rapid addition of vasopressin to the bath, P_f did not increase significantly for 23 s, and then increased rapidly over 5-10 min to reach a maximum value of ~250 x 10^{-4} cm/s. The lag time (T_{lag}) after agonist addition in which P_f did not change was reduced to 11 s when 8-bromo-cAMP

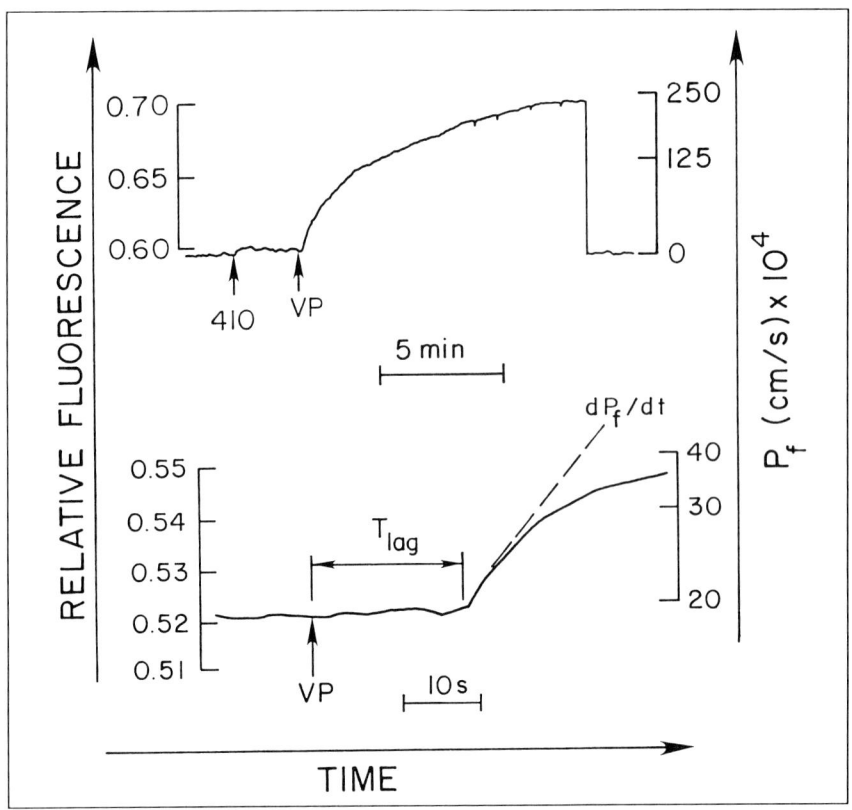

Fig. 3.1. Time course of transepithelial osmotic water permeability in isolated perfused cortical collecting duct from rabbit kidney in response to vasopressin (VP, 250 mU/ml) addition. Tubules were perfused with an isosmotic buffer (290 mOsm) containing a membrane-impermeant fluorescent indicator. Where indicated, bath osmolality was increased from 290 to 410 mOsm and vasopressin was added. Transepithelial osmotic water permeability (P_f) was measured in 1-s time intervals as described in the text. The curve is shown at the bottom on an expanded time scale to indicate the lag time (T_{lag}) between vasopressin addition and increased P_f, and the initial rate (dP_f/dt) of increase in P_f. (Adapted from ref. 135.)

was used (instead of vasopressin) as agonist. In response to removal of vasopressin from the bath, P_f decreased to baseline level over ~30 min; removal of 8-bromo-cAMP reduced the time to reach baseline level to ~5 min.

These kinetic results for the pre-steady-state turn-on and turn-off of water permeability can be interpreted in terms of the membrane shuttle hypothesis for the vasopressin hydroosmotic response. (See Fig. 1.3, Chapter 1.) According to the hypothesis, apical plasma membrane water permeability is low and basolateral membrane water permeability is high in the unstimulated collecting duct.[66,215,229] Vasopressin binding to V_2 receptors at the cell basolateral membrane causes activation of adenylate cyclase through a G_s protein. The generated cAMP activates a cAMP-dependent protein kinase A and causes phosphorylation of cytosolic proteins which may be involved in cytoskeletal regulation. Then by a series of unknown events, intracellular vesicles containing water channels fuse with the apical plasma membrane. Apical membrane water permeability is increased, giving increased transepithelial water permeability. In response to withdrawal of the vasopressin stimulus, it is thought that functional water channels are retrieved from the apical plasma membrane by an endocytic mechanism for storage or degradation. Endocytosis of water channels occurs via clathrin-coated pits in mammalian kidney.[30,217] It is remarkable that these complex biochemical and trafficking events occur in seconds and minutes. Although the cAMP-protein kinase A activation cascade is probably the primary signaling mechanism for vasopressin-induced hydroosmosis, there are important inputs by protein kinase C and calcium; in addition, there are also complex modulatory effects by many agents and signaling mechanisms including adrenergic and cholinergic agents, compounds that activate or inhibit G proteins, the arachidonic acid pathway, protein isoprenylation, and others. A discussion of these regulatory mechanisms is beyond the scope of this chapter.

WATER PERMEABILITY IN ENDOSOMES FROM KIDNEY COLLECTING DUCT

An important prediction of the membrane cycling hypothesis is that endocytic vesicles formed in response to vasopressin action should contain functional water channels. This prediction was first tested in kidney collecting duct.[257]

The Brattleboro rat was used in order to eliminate the confounding effects of endogenous vasopressin. Brattleboro rats are vasopressin-deficient but have kidneys that respond normally to exogenously-added vasopressin. Apical endocytic vesicles were labeled in vivo with membrane-impermeant fluid-phase markers (6-carboxyfluorescein or FITC-dextran) by intravenous infusion (see Fig. 3.2 for schematic). The markers were filtered by kidney glomerulus and present in the tubule lumen for fluid-phase apical endocytosis. Morphological studies of thin fixed kidney sections showed that the

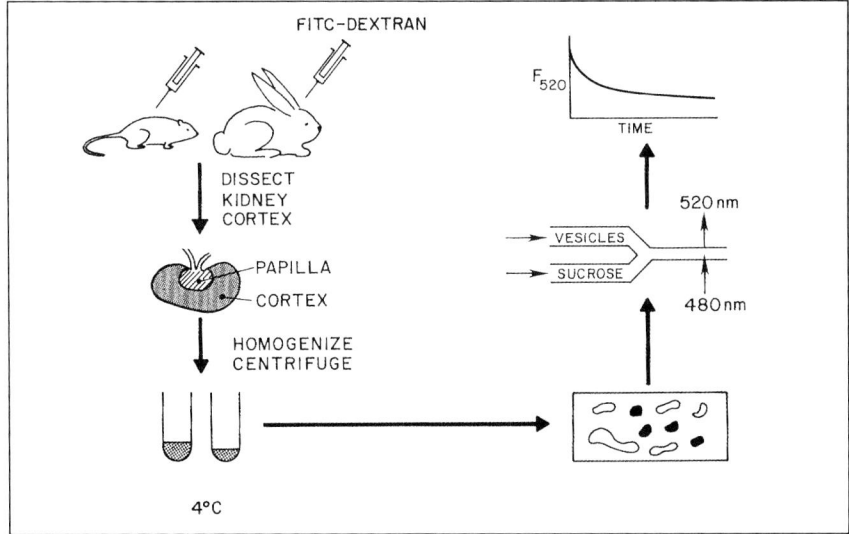

Fig. 3.2. Schematic of method for in vivo labeling of endosomes from mammalian kidney. Animals were infused intravenously with a fluid-phase marker of endocytosis, e.g., FITC-dextran. After glomerular filtration and endocytic uptake, kidneys were removed, dissected, homogenized and centrifuged to yield a microsomal pellet. The pellet contained fluorescently-labeled endosomes. Osmotic water permeability was measured in the crude microsomal pellet by stopped-flow fluorescence quenching.

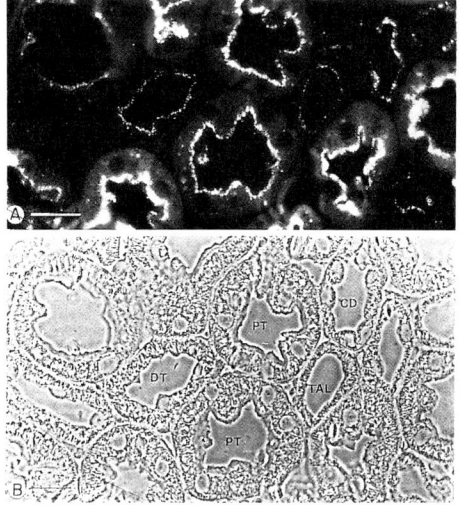

Fig. 3.3. Fluorescent labeling of endocytic vesicles from rat kidney cortex. Rats were infused with FITC-dextran, kidneys were fixed in situ, and thin sections were examined by fluorescence microscopy (top). Tubule segment identification is shown in the brightfield section (bottom). PT - proximal tubule, CD - collecting duct, TAL - thick ascending limb of Henle. (Adapted from ref. 141.)

majority of fluorescent endosomes in kidney cortex were located near the apical membrane of proximal tubule (Fig. 3.3) and the majority of fluorescent endosomes in papilla were located near the apical membrane of collecting duct principal cells;[141] the number of labeled vesicles in papilla increased with vasopressin stimulation. The kidney cortex and inner papilla were then homogenized and differentially centrifuged to produce a crude microsomal pellet. The pellet contained fluorescently labeled endosomes that had formed between the time of intravenous infusion of the fluorescent marker and animal death. Osmotic water permeability was measured in the fluorescently-labeled endosomes by a stopped-flow fluorescence quenching technique as described in Chapter 2. The majority of vesicles in the microsomal pellet, which were nonfluorescent, were not observed in the water transport measurement.

Figure 3.4 shows the time course of fluorescence quenching when a suspension of microsomes was subjected to an inwardly directed osmotic gradient.[140] The curves are shown on different time scales on the left and right to visualize the full osmotic response. Water transport in endosomes from kidney cortex was similar for the control (-VP) and vasopressin-treated (+VP) rats. In kidney outer and inner papilla, the time course of fluorescence in endosomes from control rats was fitted well to a single exponential function with a time constant of ~1 s. In the vasopressin-treated rats, a biexponential fit was required, indicating the presence of at least two populations of labeled endosomes. In addition to a population of endosomes with low-water permeability (1 s time constant), a second population of papillary endosomes had rapid water efflux (40 ms time constant). P_f was 0.03 cm/s and E_a was 3.8 kcal/mol in the water permeable endosomes from kidney papilla; in contrast, P_f was ~0.001 cm/s and E_a was 13 kcal/mol in the water impermeable endosomes.[257] Control studies were performed to show that the appearance of functional water channels in papillary endosomes was a direct action of vasopressin, and not an indirect effect of urine osmolality, fluorophore concentration, vascular changes or endosome geometry. Additional studies indicated that the number of papillary endosomes with high-water permeability correlated with the physiological state of antidiuresis.[139] The results provided direct evidence for a vasopressin-sensitive population of subcellular endosomes containing functional water channels in support of the membrane shuttle hypothesis.

PERMEABILITY OF ENDOSOMES FROM TOAD BLADDER

The toad urinary bladder served as another model system to study endosome function (Fig. 3.5). Unlike the collecting duct, it is possible to have direct control over agonist concentration and osmolalities in toad bladder. To label bladders with membrane-impermeant fluorescent markers in vitro, isolated washed bladder sacs were filled with the solutions containing fluorescent fluid-phase markers and incubated for 5-30 min under specified conditions. Bladders were washed, and epithelial cells were scraped, homogenized and centrifuged to yield a microsomal pellet. The microsomal pellet

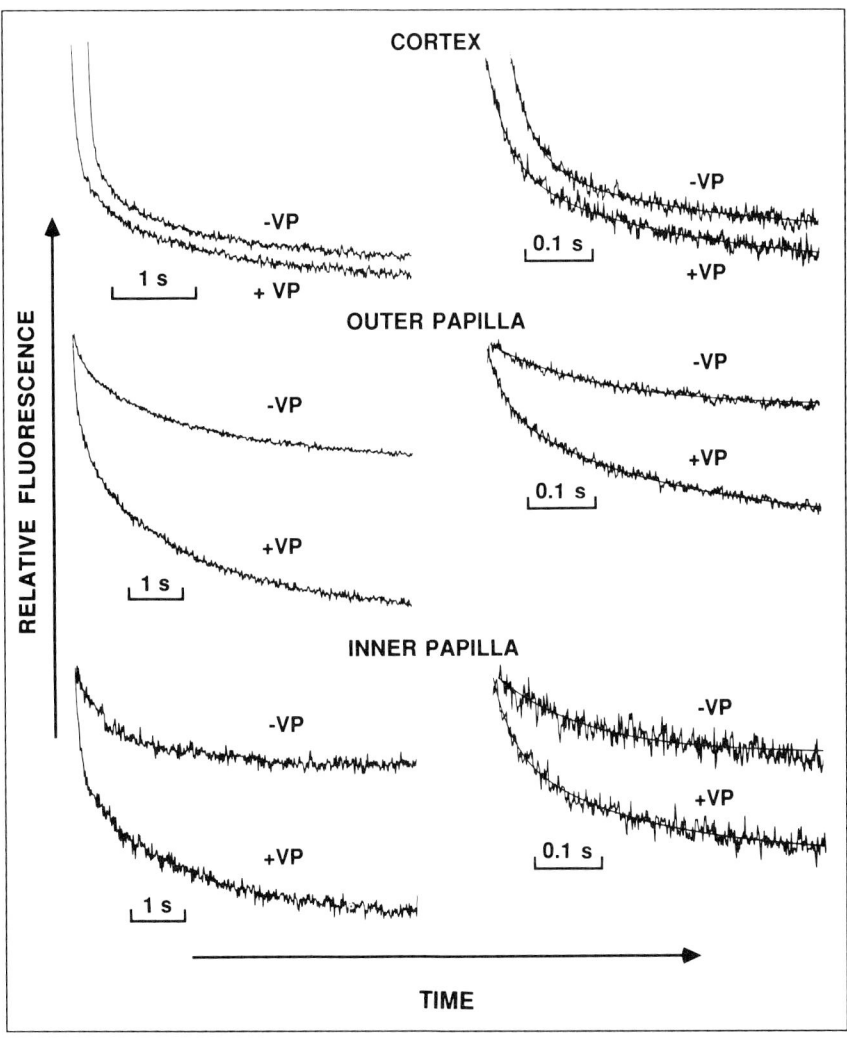

Fig. 3.4. Osmotic water permeability in endosomes from control (-VP) and vasopressin-treated (+VP) Brattleboro rats. Endosomes were labeled in vivo with carboxyfluorescein and microsomal pellets were prepared from dissected kidney cortex, outer papilla and inner papilla. Osmotic water transport was measured by the stopped-flow fluorescence quenching assay in the microsomal pellet containing the fluorescently-labeled endosomes. Data are shown on two time scales (left and right). Note the rapid downward deflection in papillary endosomes from the vasopressin-treated rats. (Adapted from ref. 140.)

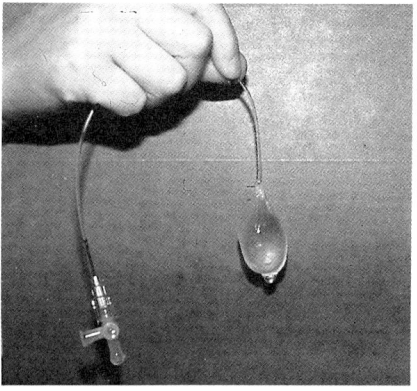

Fig. 3.5. Toad bladder model for study of vasopressin-regulated water permeability showing the toad Bufo marinus (left) and an isolated hemibladder sac (right).

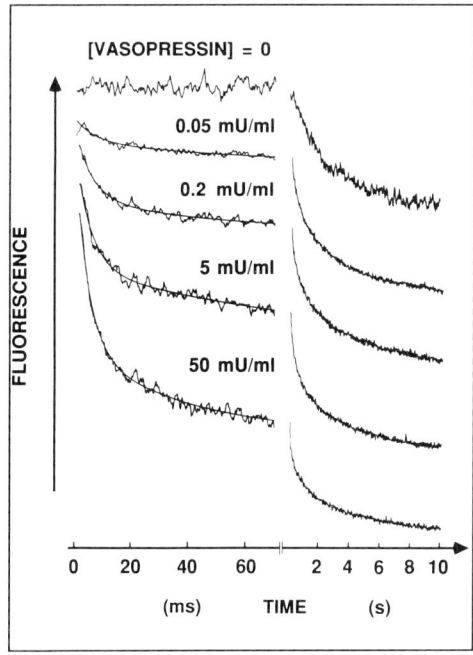

Fig. 3.6. Osmotic water permeability in fluorescently-labeled endosomes from toad urinary bladder. Isolated hemibladder sacs were bathed in toad Ringer's buffer containing the indicated concentrations of vasopressin and the lumen was labeled with isosmotic buffer containing carboxyfluorescein. Bladder cells lining the lumen were scraped, homogenized and centrifuged to yield a microsomal pellet containing the fluorescently-labeled endosomes. Osmotic water permeability was measured by the stopped-flow fluorescence quenching technique. Note the rapid downward deflection in endosomes from vasopressin-treated bladders. (Adapted from ref. 140.)

contained fluorescently labeled endosomes that were formed during the incubation with luminal fluorophore. The suspended pellet was suitable for measurements of water, solute and urea transport, as well as ATP-dependent acidification.

It was found that endosomes from toad urinary bladder contained functional water channels only when vasopressin was present at the time of fluorescence labeling.[201,204] In endosomes from vasopressin-treated bladders, fluorescence decreased very rapidly (5-10 ms) in a subpopulation of labeled vesicles in response to an inwardly directed gradient (Fig. 3.6, left); the number of vesicles with high-water permeability increased in bladders exposed to increasing concentrations of vasopressin at the time of fluorescence labeling. In addition, there was a constitutive population of labeled vesicles with low-water permeability (Fig. 3.6, right). The calculated P_f of the endosomes with high-water permeability was >0.1 cm/s, representing the highest osmotic water permeability reported in any artificial or biological membrane. The high-water permeability was >80% inhibited by HgCl$_2$ and was weakly temperature-dependent (E$_a$ 3.9 kcal/mol).[204] Water permeability was low in untreated bladders. Water permeable endosomes were absent in control studies in which bladders were exposed to vasopressin and endocytosis was inhibited by low temperature, azide incubation or glutaraldehyde fixation. These results indicated that vasopressin induces the formation of water-permeable endosomes containing functional water channels. Similar results were reported by Harris, Zeidel and coworkers[94,95,288] and by a Russian group.[129]

A series of studies were carried out to examine the number and water permeability of endosomes formed in bladders exposed to different osmolalities and agonists.[201] When bladders were fluorescently labeled in the presence of a serosal-to-mucosal osmotic gradient, the water permeability of individual endosomes decreased. However, when bladders were labeled in the absence of a transepithelial osmotic gradient, the number of labeled endosomes decreased but the water permeability of individual endosomes did not change. These results suggested that the "packing" of endosomes with water channels was sensitive to osmotic gradients; it was proposed that osmotic water flow from lumen to bath might induce endocytosis prior to complete packing (see studies on collecting duct below). Recent studies show that water channel endocytosis in toad bladder is stimulated also by cell swelling and/or cytoplasmic dilution.[94]

Experiments were performed to determine the factors that affected the formation and water permeability of endosomes from toad bladder.[205] There was no difference in the number and water permeability of endosomes formed in bladders treated with high concentrations of the cAMP agonists vasopressin, forskolin or 8-bromo cAMP. Activation of protein kinase C by bladder treatment with phorbol esters induced the formation of a small number of water permeable endosomes; interestingly, phorbol esters increased by three-fold the number and water permeability of endosomes induced by cAMP activation. In isolated endosomes containing vasopressin-sensitive water channels, water permeability was not influenced by a series of biochemical maneuvers including phosphorylation, dephosphorylation, G-protein activation, ATP, and calcium. These results suggested that vasopressin regulates water permeability in target epithelial cells by the exocytosis and endocytosis of constitutively functional water channels.

Nonelectrolyte and passive proton permeability measurements were performed to determine the selectivity of the vasopressin-sensitive water channel in toad bladder.[201] Urea permeability is stimulated by vasopressin in toad urinary bladder and inner medullary collecting duct of mammalian kidney.[269] Vasopressin stimulation of urea permeability might result from direct posttranslational modification (phosphorylation) of urea transporters in situ in the plasma membrane or membrane cycling of urea transporters together with or separate from water transporters. Urea permeability was measured in fluorescently-labeled endosomes isolated from vasopressin-treated bladders by stopped-flow fluorescence quenching. Urea permeability was low (10^{-6} cm/s) and not inhibited by phloretin and high-affinity urea analogs.[201] The vasopressin-sensitive water channel thus excludes urea and is more like a narrow channel than a wide pore. These results also indicate that vasopressin-sensitive urea and water transporters are separate proteins, consistent with recent data suggesting that the cAMP-stimulated urea transporter from kidney papilla can be activated by in situ phosphorylation.[107]

Passive proton permeability[250] was measured from the rate of pH decrease in voltage-clamped endosomes in response to a 1 pH unit gradient. Proton permeability was relatively low (0.051 cm/s) and not different from that measured in water-impermeable endosomes isolated in the absence of vasopressin.[201] Similar results were reported for water-permeable endosomes from kidney collecting duct.[140] Although there may be a finite proton conductance associated with a water channel, it was below the level of detection in these studies. The artificial gramicidin and amphotericin water channels have high-proton conductance,[109] whereas the erythrocyte water channel does not conduct protons.[149] From the physiological perspective, a high-proton conductance in toad bladder or collecting duct would be deleterious to cell function because the urine bathing the cell apical membrane is very acidic (pH <5). However, two studies have suggested that the vasopressin-sensitive water channel does have a significant proton conductance. Harris et al[97] reported a proton permeability similar to that reported by Shi et al,[201] however they concluded that the water channel was proton permeable based on inhibition of proton permeability by pCMBS. Harvey et al[102] examined the oxytocin and pH dependences of voltage-driven proton transport in frog urinary bladder and concluded that apical and basolateral water channels were proton permeable. Resolution of this issue will require measurements of proton permeability in proteoliposomes reconstituted with purified vasopressin-sensitive water channels.

Water Permeability in Toad Bladder Granules

Granular cells in amphibian urinary bladder make up ~90% of the mucosal surface and are responsible for the vasopressin hydroosmotic response. In response to stimulation by vasopressin, both aggrephores and granules fuse with the apical membrane.[87,154,267] Although aggrephores contain intramembrane particles that are believed to represent water channels, the physiological role of granule exocytosis has been unclear. To examine whether granules have a role in the vasopressin hydroosmotic response, the water permeability and protein content of purified granules was measured.[259] By stopped-flow light scattering, granules shrunk slowly (~2 s) in response to an osmotic gradient with a calculated P_f of 5 x 10^{-4} cm/s at 23°C. P_f was not inhibited by mercurials and was strongly temperature dependent (E_a 17.6 kcal/mol), which would indicate absence of functional water channels. SDS-PAGE of integral granule membrane proteins (isolated by Triton X-114 extraction) revealed several proteins in the 14-160 kDa size range, including distinct bands at 17 and 55 kDa. Because granules are water impermeable, it is unlikely that granule membrane proteins are components of the vasopressin-sensitive water channel. P_f was high in isolated surface membrane vesicles and microsomes that were prepared in parallel with the granules. Diphenylhexatriene anisotropy measurements suggested that the very low water permeability in granules was related to low membrane fluidity.

Novel Trafficking Mechanism for Vasopressin-Sensitive Water Channels

The conventional endosomal pathway in mammalian cells includes trafficking to acidic lysosomes, although a few examples exist (e.g., transferrin receptor) where recycling occurs from an early, mildly acidic compartment.[157] There have been no examples of endosomal targeting to a nonacidic compartment. However there is now evidence in both kidney collecting duct[140,194] and toad urinary bladder[41,271,288] that endosomes containing the vasopressin-sensitive water channel are nonacidic.

To determine whether water-permeable endocytic vesicles contained functional proton pumps, ATP-dependent acidification was measured in FITC-dextran-labeled endocytic vesicles from vasopressin-treated Brattleboro rats.[140] Whereas strong ATP-dependent acidification (>1 pH unit) was detected in endosomes from kidney cortex, no detectable acidification was found in endosomes from kidney papilla that contained vasopressin-sensitive water channels. The lack of ATP-dependent acidification was not due to high passive proton permeability. Morphological studies showed that endosome FITC-dextran fluorescence did not colocalize with antibodies against the lysosomal glycoprotein LGP-120 (Fig. 3.7). These results indicated that vesicles which internalize water channels in collecting duct do not contain functional proton pumps. Further, Western Blot analysis using polyclonal and monoclonal antibodies against the vacuolar-type H+-ATPase suggested that the lack of ATP-dependent acidification is due to absence of selected subunits of the vacuolar proton ATPase.[194] The 16, 31 and 70 kDa ATPase subunits were absent in immunoblots of purified endosomes using specific antibodies, whereas the 56 kDa subunit was present.

In toad urinary bladder, similar studies of ATP-dependent acidification indicated that endosomes containing water channels do not contain functional proton pumps.[271] Independent analysis of vesicle pH by DAMP immunocytochemistry indicated that nonacidic endosomes participate in recycling of vasopressin-sensitive water channels in toad bladder.[41] These results suggest that vasopressin-sensitive water channels are stored in a novel nonacidic compartment after endocytic retrieval. The factor(s) responsible for targeting to this compartment are not known, nor is it known whether internalized water channels eventually recycle to the apical plasma membrane upon rechallenge with vasopressin. Some of the newer imaging techniques[123,290] should facilitate the direct measurement of pH in intact vasopressin-responsive cells, and the tracking of individual fluorescently-labeled endosomes.

Taken together, the functional data suggest that vasopressin-sensitive water transporters are cycled to and from the apical plasma membrane of target epithelial cells (Fig. 3.8). The vasopressin-sensitive water channel probably excludes urea and protons, suggesting separate mechanisms for regulation of water and urea permeability. Water channels are retrieved into a nonacidic compartment without trafficking to lysosomes.

REGULATION OF VESICULAR TRAFFICKING BY OSMOTIC GRADIENTS

Several studies suggested that the presence of a serosal-to-mucosal osmotic gradient (serosal > mucosal osmolality) influenced endocytosis in toad urinary bladder.[100,153] A serosal-to-mucosal gradient increased the endocytic retrieval of fluid-phase markers and enhanced the down-regulation of water permeability in response to prolonged vasopressin stimulation.

To investigate whether similar effects occur in kidney collecting duct and to elucidate the mechanism by which osmotic gradients influence endocytosis, P_f measurements were made in kidney collecting duct.[133] P_f was measured continuously by the luminal fluorescence technique.[131] The kinetics of turn-on and turn-off of water permeability were measured in the presence of various bath and lumen osmolalities. The principle observations were that bath > lumen osmolality decreased vasopressin-dependent turn-on

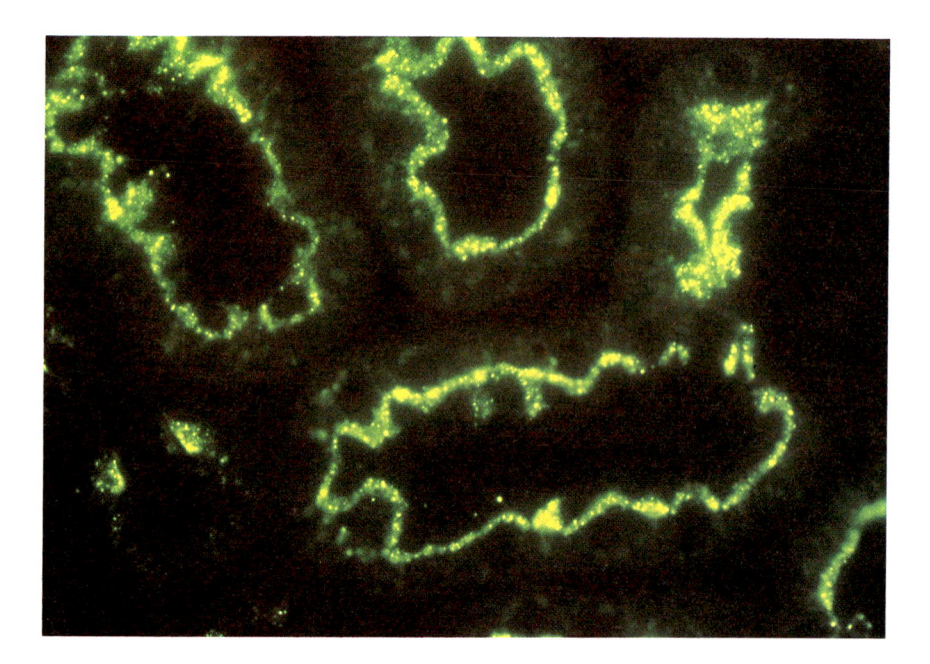

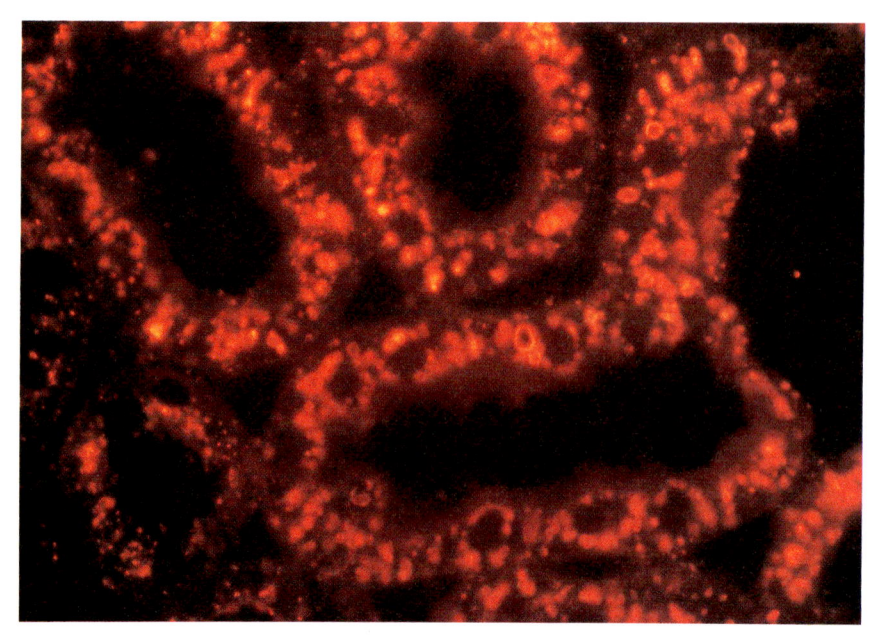

Fig. 3.7. Separate locations for vesicles containing vasopressin-sensitive water channels and lysosomes in kidney inner medullary collecting duct of vasopressin-treated Brattleboro rat. Endosomes were labeled in vivo with FITC-dextran. FITC-labeled endosomes (green, top) did not colocalize with lysosomal marker LGP120 stained with specific antibody and visualized with a rhodamine-labeled secondary antibody (red, bottom). (Adapted from ref. 140.)

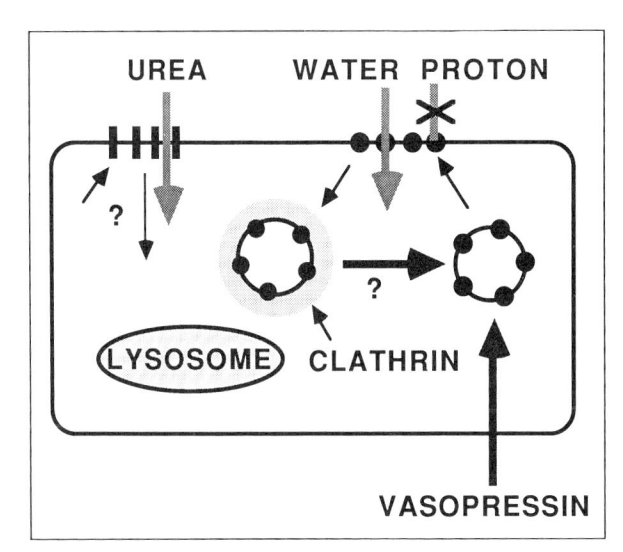

Fig. 3.8. Membrane shuttling of vasopressin-sensitive water channels showing endocytosis into a non-acidic vesicular compartment without lysosomal fusion, and exclusion of urea and protons from the water pathway. See text for details.

and increased turn-off of water permeability, whereas lumen > bath osmolality increased vasopressin-dependent turn-on and decreased turn-off of water permeability (Fig. 3.9). Steady-state P_f in the presence of vasopressin was not sensitive to osmotic gradient size or direction. Control studies showed that these effects were not caused by paracellular water transport, asymmetric transcellular water permeability, or changes in cell volume. In addition, vasopressin-dependent endocytosis of luminal rhodamine-dextran in collecting duct was increased by bath > lumen osmolalities and decreased by lumen > bath osmolalities. The results indicated that transcellular volume flow modulates vasopressin-dependent cycling of vesicles containing water channels, suggesting a novel driving mechanism to aid or oppose the targeted, hormonally-directed movement of subcellular membranes.

Relationship Between Water Permeability and Cell Fluidity

In early studies, it was proposed that the vasopressin hydroosmotic response was mediated in part by changes in apical plasma membrane fluidity and/or subapical cytoplasmic viscosity. Grantham showed that vasopressin increased the deformability of the apical membrane of cortical collecting duct.[83] Pietras and Wright showed that vasopressin increased nonspecifically the permeability of toad bladder to small nonelectrolytes.[180] Using pyrelene as a probe of membrane fluidity in isolated toad bladder cells, Masters showed that cAMP caused a small increase in fluidity that paralleled the predicted change in water permeability.[152] Le Grimellec and coworkers used trimethylammonium diphenylhexatriene (TMA-DPH) anisotropy to show differences in lipid order between apical and basolateral membranes in MDCK cells, and to show small changes in membrane order with vasopressin.[77,78] There is a considerable body of evidence suggesting a relationship between membrane fluidity and water permeability.

To test the hypothesis that tubule cell membrane fluidity is influenced by vasopressin, an "anisotropy imaging" technique was developed to map fluidity in microscopic samples.[72] Steady-state fluorescence anisotropy was measured from the ratio of background-subtracted images obtained with polarized excitation light and parallel and perpendicular orientations of an analyzing emission polarizer. This approach was used to measure anisotropy in aqueous compartments in which fluorophore orientation is random. The measurement of anisotropy in nonrandomly orientated fluorophores is more difficult because the apparent anisotropy depends both on the "real" anisotropy (anisotropy measured in the hypothetical situation that fluorophore

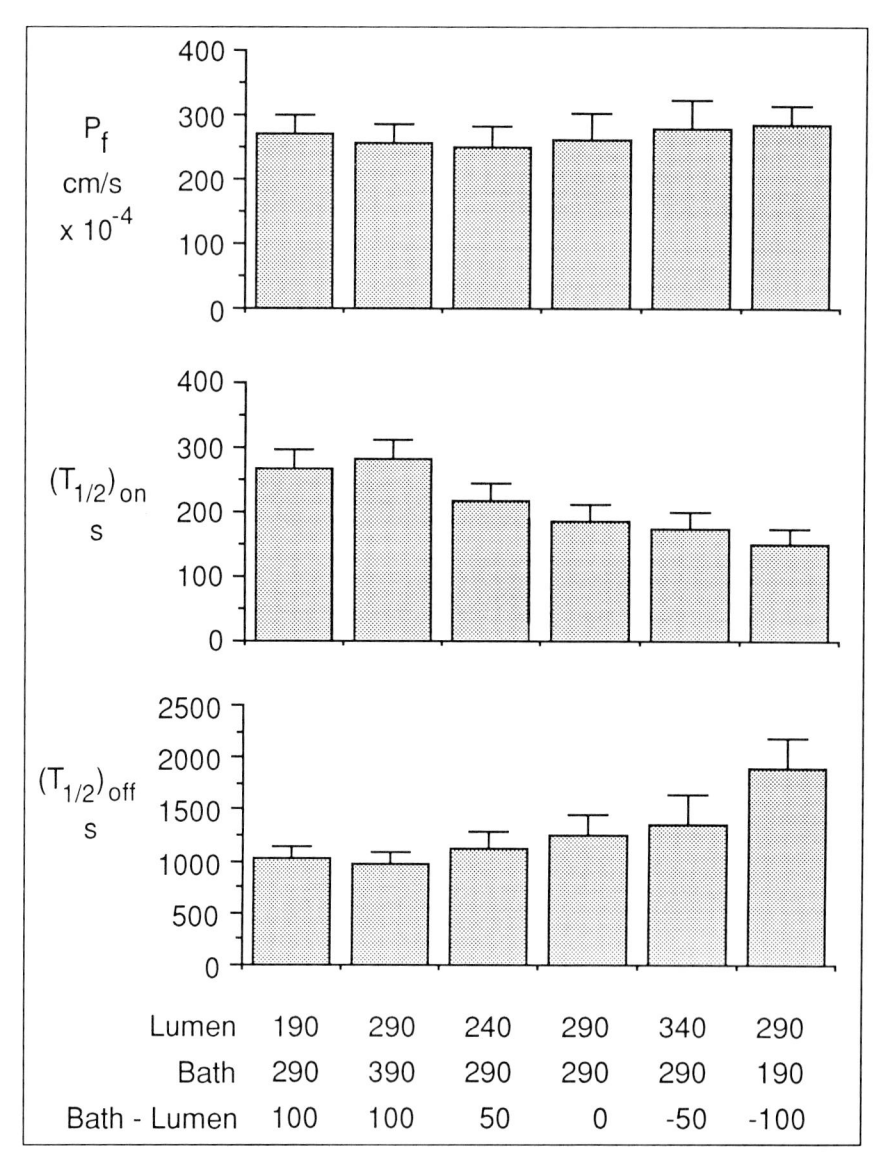

Fig. 3.9. Vasopressin-stimulated transepithelial osmotic water permeability in isolated perfused cortical collecting ducts from rabbit kidney. Tubules were perfused and bathed in buffers of indicated osmolalities and P_f was measured continuously from the fluorescence of a luminal marker. The ordinates indicate steady-state P_f in the presence of 250 mU/ml vasopressin (top), the half-time for turn-on of P_f after vasopressin addition (middle), and the half-time for turn-off of P_f after vasopressin removal (bottom). See text for explanations. (Adapted from ref. 133.)

orientation is random) and on fluorophore orientation. For the special case of a kidney tubule in which fluorophores are arranged around a cylinder, a mathematical model was developed to place narrow limits on true anisotropy from measured anisotropy.[72,75] Experimentally, the apical plasma membrane of isolated perfused collecting duct was stained by perfusion with the fluidity-sensitive fluorophore TMA-DPH. The tubule was perfused on the stage of an epifluorescence microscope with polarized excitation light (365 nm) and a rotatable analyzing polarizer. Images were acquired before and after staining at a series of angles of the analyzing polarizer. Anisotropy was calculated by an "orientation-independent" method as described in detail in the study of Fushimi et al.[72]

In cortical collecting duct from rabbit kidney, TMA-DPH anisotropy was 0.254 in apical membrane, similar to that of 0.252 in basolateral membrane.[75] Serosal vasopressin at a concentration that increased P_f >10-fold did not affect anisotropy significantly. Control experiments involving added membrane fluiding agents and temperature effects showed that anisotropy was sensitive to changes in membrane fluidity; the upper limit to the change in apical membrane fluidity caused by vasopressin was less than that caused by a 2°C change in temperature. Therefore, although changes in fluidity in localized membrane compartments cannot be ruled out, these results indicated that changes in bulk membrane fluidity do not accompany vasopressin-induced hydroosmosis. Measurements of membrane fluidity in isolated perfused proximal tubule and ascending limb of Henle revealed no correlation between water permeability and fluidity.[72,75]

To test the hypothesis that the rheology of subapical membrane cytoplasm is modified by vasopressin action, a picosecond microscopy technique was developed to quantify fluid-phase viscosity in living cells.[74,122,176,247] Cell cytoplasm was stained by an aqueous phase fluorophore (e.g., BCECF). Picosecond fluorophore rotation was measured by multi-harmonic frequency-domain microfluorimetry.[252] Cells were illuminated with impulse-modulated polarized monochromatic light from an argon laser (Fig. 3.10). Reference light and emitted fluorescence were detected by phase-sensitive photomultipliers using a Fourier transform cross-correlation detection method. Measurements were carried out with parallel and perpendicular orientations of the analyzing polarizer to calculate differential phase angles and modulation ratios. The time-domain rotational parameters describing anisotropy decay were fitted by nonlinear regression. In Swiss 3T3 fibroblasts, fluid-phase viscosity was only ~30% greater than that of water and not sensitive to cell volume, stage of cell cycle, or second messengers.[74] There was no evidence for the existence of "organized water" in cell cytoplasm. In contrast, fluorophore translational diffusion was slowed ~4-fold compared to that in water.[122] The primary factor responsible for the slowed translational diffusion was collisional interactions with mobile and immobile components in cell cytoplasm.

Picosecond rotational measurements were carried out in isolated kidney tubules in which cell cytoplasm was loaded with BCECF. In rabbit proximal straight tubule, ascending limb of Henle, and cortical collecting duct, fluid-phase viscosity was only 10-50% greater than that of water;[177] there was no effect of vasopressin on fluid-phase viscosity in collecting duct. Interestingly, BCECF translation and rotation in MDCK cells grown in hypertonic media was increased significantly because of accumulation of organic osmolytes.[177] These results suggest that changes in fluid-phase cytoplasmic rheology do not accompany the vasopressin hydroosmotic response. Novel internal reflection methods[18] will be required to determine whether the properties of subapical cytoplasm very near the membrane are sensitive to hormone stimulation.

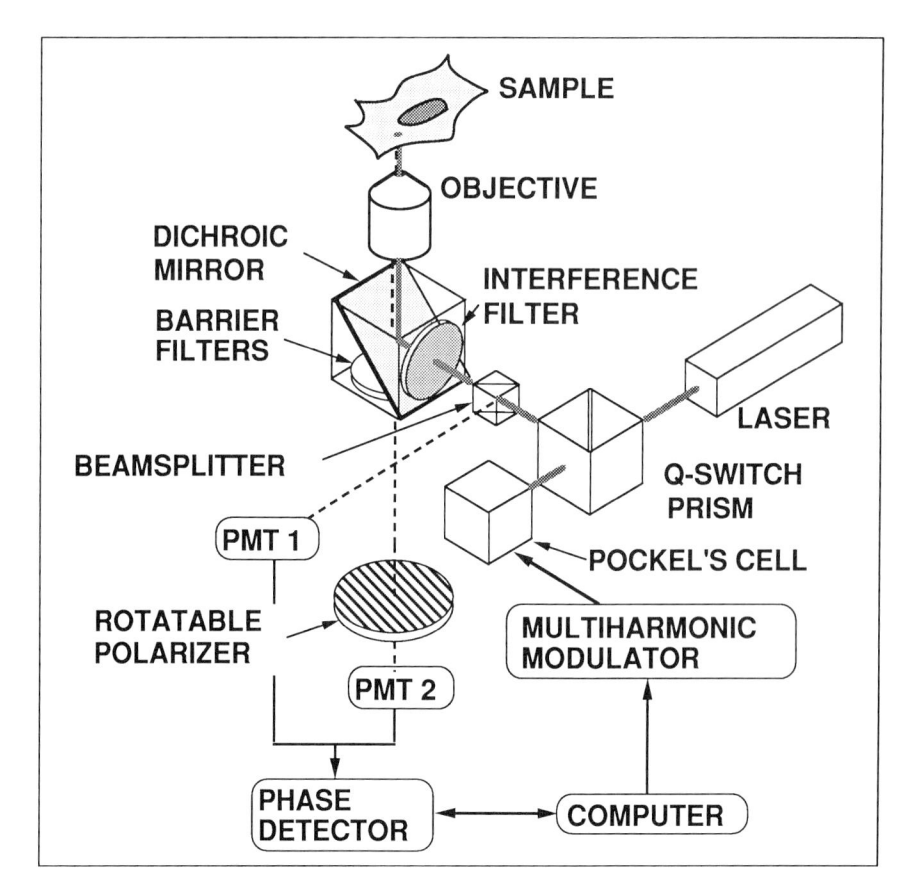

Fig. 3.10. Instrumentation for measurement of fluid-phase cytoplasmic viscosity. A cell or tubule sample was excited with a focused beam of polarized, impulse-modulated light from a laser. Multiharmonic modulation was obtained by a high-frequency generator/ pulse shaper, Pockel's cell and Q-switch prism. The modulated beam was split into reference and sample beams. The sample beam was focused onto the cell; emitted fluorescence was collected, filtered and passed through an analyzing polarizer. Fluorescence was detected by a cross-correlation method using gain-modulated photomultipliers.[252] Multifrequency phase angles and modulation factors were determined by Fourier transform. (Adapted from ref. 74.)

CHAPTER 4

THE SEARCH FOR WATER CHANNELS

The existence of specific water transporting proteins has been proposed for many years.[50,258,262] Special difficulties made the identification of water transporters particularly challenging, including: a) the lack of selective water transport inhibitors to label water channels, b) the lack of ligand protection methods, c) the relatively high-basal water permeability of lipid bilayers not containing water channels, and d) the technical difficulties in the measurement of water permeability. In addition, the possibility that lipids comprise part of the water transporting pathway made it uncertain whether conventional protein purification and molecular cloning methods would be successful. Finally, the ability of a number of proteins to pass some water confounded the search for a selective water transporting protein. This chapter reviews the biophysical evidence for water transporting proteins in certain cell membranes and biochemical and antibody attempts to label water channels. The expression of water channels in *Xenopus* oocytes from tissue mRNA then provided evidence for the existence of distinct water transporting proteins, and radiation inactivation measurements provided a molecular size for the putative water transporter. These observation lead to the recognition that CHIP28, an erythrocyte protein isolated and cloned without a known function, was indeed the erythrocyte water channel and the first member of a family of water channels.

WATER TRANSPORT IN ERYTHROCYTES

The erythrocyte has been the most extensively studied water-transporting cell. Much of the early work was performed by Solomon and coworkers and Macey and coworkers who showed that erythrocyte P_f was high (~0.02 cm/s at 20°C) and 90% inhibited by organic mercurials such as p-chloromercuribenzene sulfonate (pCMBS).[80,149,212,213] The activation energy (E_a) for P_f was low (4.5 kcal/mol) in the absence of inhibitors, and became high (~10 kcal/mol) after inhibition by mercurials. The residual P_f after inhibition is probably mediated by membrane lipids. Erythrocytes from virtually all species have high-water permeability; rabbit erythrocytes were found recently to have an exceptionally high P_f of 0.05 cm/s that was inhibited by 98% by mercurials.[222] Based on original P_d estimates by ^3H_2O exchange, erythrocyte P_f/P_d was calculated to be 3.4;[212] taken together with relative erythrocyte permeabilities for a series of polar nonelectrolytes of different sizes, the existence of a wide pore with ~0.9 nm diameter was proposed.[80] The determination of a low reflection coefficient for urea of ~0.6 provided further support for an aqueous channel that passed water and urea. A morphological study of membrane intercalated particles in erythrocyte membranes provided further support for the existence of water pores.[44] These early results clearly established the existence of a "facilitated" transporting pathway in erythrocytes.

Although the existence of an erythrocyte water channel is now established unequivocally, the early conclusions about transporter geometry and specificity were not correct. More recent estimates give an erythrocyte P_f/P_d of 11;[163] the difference from earlier values was related to improved measurement accuracy and the subtraction of nonchannel mediated transport. Although measurement of the urea reflection coefficient in erythrocytes is challenging because of the high urea permeability, recent data suggests that its value is near unity.[143] It was pointed out by Macey that passage of water and urea through a common pore was unlikely because erythrocytes from several mammalian species (including human erythrocytes with blood type $J_k[a-b-])$[70] have low urea and high water permeabilities.[149] The inverse relationship between nonelectrolyte permeability and size in the early studies was probably fortuitous. These more recent observations suggested that the erythrocyte water transporter is a narrow channel rather than a wide pore. As described in Chapter 5, permeability measurements of erythrocyte CHIP28 water channels in reconstitutued proteoliposomes and *Xenopus* oocytes support the conclusion that the erythrocyte water transporter does not pass urea and other solutes.

Water Transport in Kidney Proximal Tubule

There is strong evidence that kidney proximal tubule plasma membranes contain functional water channels.[246] Physiologically, the high-water permeability facilitates the near isosmotic reabsorption of glomerular filtrate. Most values for P_f in isolated perfused proximal tubules from rat and rabbit were in the range 0.1-0.5 cm/s, although values of 0.04 and 2.2 have been reported (see ref. 15 for review). Some of the variability has been attributed to technical factors, including luminal perfusion rate, tubule viability, and the permeabilities and reflection coefficients of luminal and bath solutes. A more recent study suggested that much of the variability can be accounted for by differences in osmotic gradient size.[16] P_f in rabbit proximal tubule increased from 0.08 to 0.5 cm/s as transepithelial osmotic gradient size decreased from 100 to 20 mOsm. By examination of the effects of inhibitors, temperature and external viscosity, it was concluded that the decrease in P_f with increasing osmotic gradient size was probably due to a flow-dependent intracellular unstirred layer. However, these in vitro studies performed with relatively large osmotic gradients must be viewed with caution[278] because very small osmotic gradients exist across proximal tubule membranes in vivo.

The high-water permeability in proximal tubule could in principal be accomplished by high transcellular and/or paracellular water permeabilities. The observation that the mercurial pCMBS inhibited water permeability in rabbit proximal tubule by >80% supported the view that the majority of osmotically-induced water transport was transcellular.[16] It is unlikely that pCMBS would inhibit transport through a paracellular pathway. The conclusion that water moves primarily through proximal tubule cells was supported by theoretical estimates of paracellular water permeability based on the morphology of junctional complexes,[178] and measured solute permeabilities for a series of small polar nonelectrolytes.[183] P_f in the separate apical and basolateral membranes in intact perfused proximal tubule has been estimated by various video methods in which cell volume is measured after a sudden change in bath or lumen osmolality;[216,273] in some experiments, the bath or lumen contained oil to prevent water movement. Apical and basolateral membrane permeabilities in the range 0.15-0.6 cm/s have been measured. However these values may underestimate actual membrane permeabilities because of finite solution exchange times, intracellular solute polarization arising from large osmotic gradients, and in some experiments, the presence of a water-permeable contralateral membrane. By careful comparison of transepithelial P_f with P_f for individual apical and basolateral membranes, Whittembury and coworkers suggested that significant transport of water occurs by a paracellular route.[81,277] The significance of the paracellular pathway for water transport in proximal tubule thus remains unresolved.

A series of water permeability measurements have been carried out on purified plasma membrane vesicles from proximal tubule apical membrane (brush border vesicles) and basolateral membrane. The first light scattering measurements performed in brush border vesicles in 1985 showed a high P_f 0.012 cm/s at 37°C (Fig. 4.1, top) with a low-activation energy of 2 kcal/mol below 33°C.[253] Urea permeability was low at 2×10^{-6} cm/s (Fig. 4.1, bottom). Subsequent studies of water permeability in brush border and basolateral membrane vesicles from rat and rabbit indicated high P_f, low E_a and reversible inhibition by $HgCl_2$.[159,181,233,256] In general, P_f values were higher in rat than in rabbit, and in basolateral than in brush border vesicles. A study of axial heterogeneity revealed higher P_f in vesicles derived from deep vs. superficial rat renal cortex.[228] There was reasonable agreement between P_f measured in vesicles with P_f in the intact proximal tubule (by video methods, see above) after correction for membrane folding.[246] The vesicle studies indicated the presence of functional water channels in kidney proximal tubule plasma membranes with properties similar to those in erythrocytes.

Diffusional water permeability could not be measured in vesicles because of the rapid water exchange times of <1 ms. However, P_f and P_d could be measured in freshly suspended proximal tubule cells by stopped-flow light scattering and NMR methods, respectively.[159] P_f was 0.010 cm/s and P_d was 0.0032 cm/s at 37°C, giving P_f/P_d of 3.1. P_f was independent of osmotic gradient size and inhibited by $HgCl_2$. However, information about the permeability properties of individual apical and basolateral membranes could not be obtained by studies on suspended cells. To measure basolateral membrane P_d, NMR measurements were carried out in fresh suspensions of rabbit proximal tubules in the presence of the external paramagnetic quencher Mn.[264] Figure 4.2 shows original data for the decay of magnetization as a function of temperature. Analysis of magnetization decay by a three-compartment model gave a P_d value of 0.002 cm/s, much lower than P_f measured in basolateral membrane vesicles or intact tubules (by video methods). P_d had an activation energy of 2.9 kcal/mol and was 60%

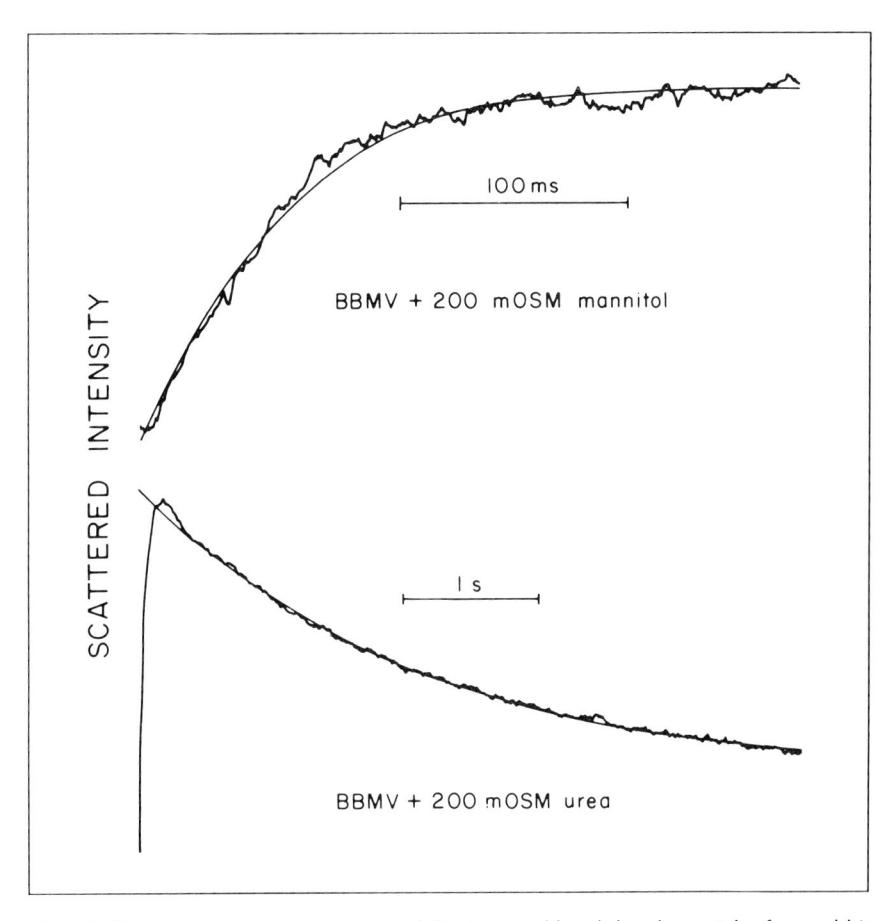

Fig. 4.1. Osmotic water and urea permeability in apical brush border vesicles from rabbit renal cortex. The time course of scattered light intensity is shown in response to 200 mM inwardly directed gradients of mannitol and urea at 37°C. Fitted curves correspond to P_f = 0.011 cm/s and P_{urea} = 2 x 10^6 cm/s. (From ref. 253.)

inhibited by 2 mM pCMBS. Taken together, these results indicate $P_f/P_d > 3$ in plasma membranes in kidney proximal tubule, supporting the presence of functional water channels.

It has been difficult to measure transepithelial P_d in isolated perfused proximal tubules by ^3H_2O diffusion because of large unstirred layers; after correction for unstirred layers by measurement of n-butanol permeability, transepithelial P_d was estimated to be 0.013 cm/s.[14] Assuming a transepithelial P_f of 0.5 cm/s (see above), transepithelial P_f/P_d was ~40. This value predicts a wide pore of unreasonably large diameter (2 nm) or a narrow channel of unreasonable length (10 nm) (see Chapter 2), suggesting that actual transepithelial P_d (after correction for unstirred layers) is much higher.

Water Channels in Subcellular Vesicles in Proximal Tubule

There is functional evidence for water channels in subcellular vesicles from kidney proximal tubule, including endocytic vesicles[283] and clathrin-coated vesicles.[263] Endocytic vesicles from rat and rabbit proximal tubule were fluorescently labeled with 6CF or FITC-dextran by intravenous injection

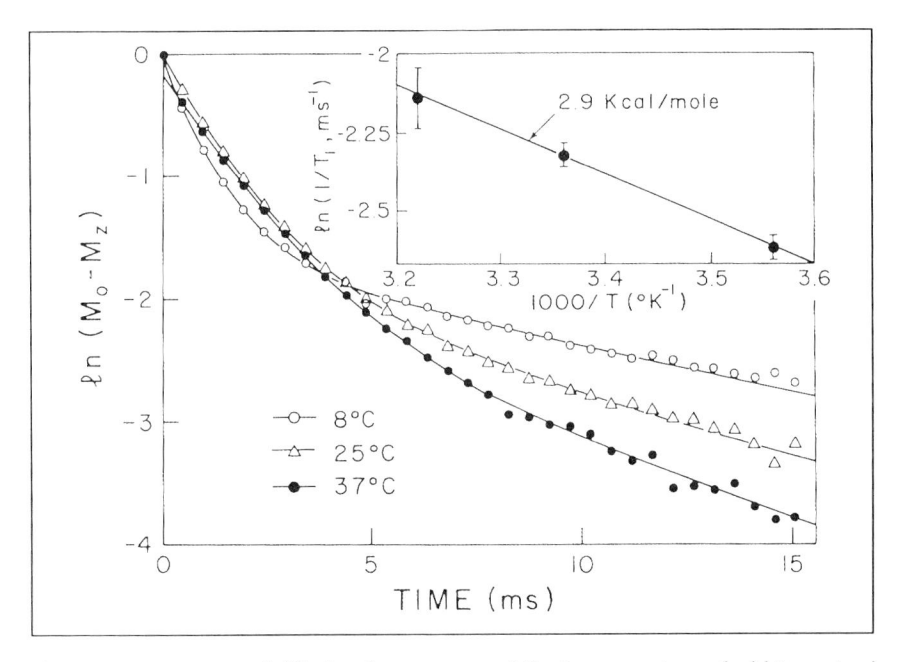

Fig. 4.2. Measurement of diffusional water permeability in suspensions of rabbit proximal tubules by proton nuclear magnetic resonance. Mn was added to the tubule suspension as a paramagnetic quencher to decrease proton relaxation times in the extracellular aqueous medium. Biexponential decay of magnetization was observed at all three temperatures. Arrhenius analysis of the relaxation times for diffusional water exchange gave an activation energy of 2.9 kcal/mol. See text for details. (From ref. 264.)

as described in Chapter 3 for the labeling of endocytic vesicles containing vasopressin-sensitive water channels. Osmotic water permeability in a crude microsomal fraction containing the labeled endocytic vesicles was measured by the stopped-flow fluorescence quenching method. Fluorescence micrographs of fixed kidney sections indicated that the majority of labeled endocytic vesicles in renal cortex were near the apical plasma membrane of proximal tubule.[141] In endocytic vesicles from rat proximal tubule, P_f was 0.029 cm/s at 23°C and reversibly inhibited by 0.2 mM $HgCl_2$; the activation energy for P_f was 6.4 kcal/mol.[283]

ATP-dependent acidification was also demonstrated in the fluorescently labeled endocytic vesicles by the decrease in pH in response to ATP addition; inhibition of ATP-dependent acidification by N-ethylmaleimide (NEM) demonstrated the presence of a vacuolar proton ATPase.[202,283] To examine whether water channels and proton pumps colocalized in the same endosomes, the time course of FITC-dextran fluorescence was measured in response to an osmotic gradient in the presence and absence of ATP.[283] The fluorescence of FITC-dextran is sensitive to both vesicle pH and volume. If water channels and proton pumps were present on separate endosome populations, then the signal amplitude arising from water channel-containing endosomes in the stopped-flow fluorescence quenching experiment would not be affected by ATP. It was found that the fluorescence amplitude decreased by 43-47% upon addition of ATP (where pH dropped by ~1 unit), indicating that nearly all labeled endocytic vesicles contained both functional water channels and proton pumps. These vesicles also contained chloride channels that are activated by protein kinase A-induced phosphorylation.[4] Light scattering measurements performed using purified endosomes from rat kidney cortex also demonstrated high water permeability that was inhibited by $HgCl_2$.[192] The retrieval of water channels from the apical membrane of proximal tubule is consistent with the high rate of apical endocytosis in proximal tubule.[190] There is no information about whether plasma membrane water channel expression in proximal tubule is regulated. The presence of water channels in an endosomal compartment, like that described for vasopressin-sensitive water channels in Chapter 3, indicates the possibility of physiological regulation.

Apical endocytosis in proximal tubule occurs by formation of clathrin-coated pits.[190] To determine whether clathrin-coated vesicles contained water channels, P_f was measured in purified clathrin-coated vesicles from bovine brain and kidney.[263] Vesicles were isolated by tissue homogenization, differential and Ficoll gradient centrifugation and Sephacryl-S100 column chromatography. In clathrin-coated vesicles from brain, P_f was low (0.002 cm/s at 23°C), strongly temperature-dependent (17 kcal/mol) and not inhibited by $HgCl_2$. In contrast, water permeability was heterogeneous in clathrin-coated vesicles from kidney. Approximately half of the vesicles isolated from kidney cortex and one-fourth of vesicles from kidney papilla had high P_f (0.02 cm/s) that was weakly temperature-dependent (2-3 kcal/mol) and inhibited by $HgCl_2$. These water permeable vesicles may originate from apical membrane endocytosis and/or vesicular transport between intracellular compartments.

SOLUTE REFLECTION COEFFICIENTS IN PROXIMAL TUBULE

As described in Chapter 2, the solute reflection coefficient provides information about the pathways for water and solute transport. The urea reflection coefficient in brush border membranes was found by light scattering and fluorescence quenching methods to be ~unity,[38] consistent with exclusion of urea from the proximal tubule water channel. The NaCl reflection coefficient was measured in brush border and basolateral vesicles from rabbit proximal tubule by both induced osmosis and solvent drag methods.[174] In the induced osmosis method, the concentration of external NaCl needed to give zero initial volume flow in sucrose-loaded vesicles was measured by stopped-flow light scattering. In the solvent drag method, the inward drag of NaCl by osmotic water movement was measured using the entrapped chloride-sensitive fluorescent indicator SPQ (6-methoxy-N-[3-sulfopropyl] quinolinium[245]). Both induced osmosis and solvent drag methods gave NaCl reflection coefficients of near unity in brush border and basolateral membrane vesicles. Initial estimates of NaCl reflection coefficients in rat brush border vesicles yielded low values of ~0.5;[182] however, subsequent measurements by the same authors gave values of unity after correction for refractive index artifacts.[230,231] The ~unity reflection coefficients for urea and NaCl suggested that the proximal tubule water transporter excludes small solutes and monovalent ions.

In intact proximal tubule, the transepithelial NaCl reflection coefficient is important to understand the mechanisms of fluid reabsorption and passive cell volume regulation.[42,85,272] In isolated perfused proximal tubules from rat and rabbit, the transepithelial NaCl reflection coefficient has been measured from the volume flow induced by equal gradients of NaCl and an impermeant solute[86] and by solvent drag.[42,117] Values in the range 0.35-1 have been reported. All measurements were made by timed collections of luminal fluid with fixed tubule length and using mean osmotic and solute gradients in the data analysis. A fluorescence approach was developed to address some of these difficulties.[203] Isolated proximal straight tubules from rabbit were perfused with buffers containing zero chloride, the chloride-sensitive indicator SPQ, and a chloride-insensitive fluorophore with a different fluorescence spectrum (fluorecein sulfonate). Tubules were bathed in buffers of a series of cryoscopic osmolalities containing NaCl. The transepithelial chloride gradient along the axis of the tubule was measured by quantitative ratio imaging. Axial chloride concentration profiles were analyzed by the Kedem-Katchalsky equations using tubule geometry, lumen flow and transepithelial P_f and P_{NaCl}. The analysis gave a transepithelial NaCl reflection coefficient of ~unity. To measure the NaCl reflection coefficient of proximal tubule basolateral membrane, the time course of NaCl influx in SPQ-loaded tubules was measured in response to rapid chloride addition to the bath in the presence of different cryoscopic osmotic gradients. The kinetic data were fitted to a nonequilibrium thermodynamic model. The basolateral NaCl reflection coefficient was also found to be near unity. Therefore, NaCl solvent drag did not occur in rabbit proximal straight tubule at 23°C.

WATER TRANSPORT IN EXTRARENAL MEMBRANES

The older literature reports a number of water transport measurements across mammalian and nonmammalian tissues.[110] However it was not possible to deduce whether facilitated transporting pathways for water were present because of unstirred layers and the absence of data about tissue geometry, mercurial inhibition and activation energy. There have been a few studies of water in isolated cells and membrane vesicles from nonrenal tissues. In human platelets, measurements of P_f by light scattering[160] and P_d by NMR[280] indicated low P_f, P_f/P_d ~1, high E_a, and no mercurial inhibition, suggesting the absence of water channels. Several studies were carried out in isolated plasma membranes from epithelial cells thought to participate in fluid movement. Biophysical measurements of water transport in brush border membrane vesicles from placenta,[114,119] trachea,[281] small intestine,[49,282] stomach[187] and unstimulated toad bladder[88] suggested the absence of water channels. Because of heterogeneity in vesicle size and composition in these studies, it is difficult to conclude whether water channels are present. As described in Chapter 8, there is now unambiguous evidence by in situ hybridization and antibody staining that specific water channels are distributed widely in fluid-transporting tissues.

BIOCHEMICAL LABELING OF PUTATIVE WATER TRANSPORTERS

There have been a number of attempts to identify putative water transporting proteins by selective protein labeling in native cell membranes. In an early study, Brown concluded that the water transport inhibitor DTNB labeled primarily the band 3 anion exchanger in human erythrocyte membranes.[32] Subsequent studies of [203]Hg-labeled pCMBS binding also showed incorporation by band 3, as well as minor labeling of a number of other bands. In a study by Solomon and coworkers,[171] low molecular weight bands (including 28 kDa) were not included in the gel; in a study by Benga et al,[11] ~3% of total pCMBS radioactivity was in the 25-30 kDa region whereas ~50% of radioactivity was associated with band 3. This observation, taken together with biophysical estimates suggesting that the erythrocyte must contain >10^5 water channels,[145,213] led to the hypothesis that band 3 was a multifunctional protein that transports anions, water and urea. However, a difficulty with these labeling studies is that many proteins contain -SH groups that are accessible to sulfhydryl-reactive reagents. It is therefore not unexpected that sulfhydryls are localized to abundant proteins such as erythrocyte band 3. As described below, expression studies in *Xenopus* oocytes indicated that band 3 is not an erythrocyte water transporter.[292]

There have been attempts to identify water channels by selective proteolysis. Erythrocyte water permeability was found however to be quite resistant to protease action.[12] In rat kidney, the sensitivity to a large series of proteases was measured in brush border vesicles (right side-out water channels) and purified endosomes (presumably inside-out water channels).[192] Whereas many of the proteases strongly inhibited the activities of ectoenzymes and integral membrane transporters, no significant effects on water transport or on the inhibition of water transport by mercurials were found. The protease resistance of the kidney proximal tubule water transporter suggested a hydrophobic protein without accessible extramembrane polar domains that are involved in water transport.

Several attempts have been made to resolve proteins that were labeled or isolated from purified apical membranes derived from vasopressin-treated amphibian urinary bladder. Harris and coworkers labeled apical membranes from toad bladder with lactoperoxidase and ^{125}I; SDS-PAGE showed proteins at 7, 14-17 and 53-55 kDa.[92] These proteins were present only in stimulated bladders. The bands at 17, 53 and 55 kDa were found to be integral membrane proteins.[101] Valenti et al used a membrane-impermeant fluorescent probe to label selectively the apical membrane of vasopressin-stimulated frog urinary bladder; SDS-PAGE of Triton-X100 extracted proteins showed labeled bands at 17, 26, 32-35 and 52 kDa.[227] These proteins were present only in extracts from stimulated bladders. Subsequently, these workers purified bladder apical membrane on charged beads—a band at 32 kDa was present on SDS-PAGE of proteins from stimulated frog bladders.[226] Verbavatz et al used a modified cryofracture technique to obtain sheets of apical membrane from frog bladder. Two-dimensional gel electrophoresis showed bands at 30-32, 44, 53, 61 and 66 kDa.[241,243] Recently, Harris et al showed that the 53 and 55 kDa bands had structural similarity and high-cysteine content.[96] Van der Goot et al used fluorescence cell sorting to enrich a population of fluorescently-labeled endosomes from toad bladder;[232] SDS-PAGE showed bands at 43, 55 and 7 kDa. The relationship between the labeled components in the studies above and vasopressin-stimulated water transport remains uncertain. In particular, bands at 17 and 55 kDa were found in purified granule membranes from toad bladder; granules are water impermeable and fuse with the apical plasma membrane upon vasopressin stimulation.[259]

Antibody Labeling of Putative Water Transproters

There have been attempts to identify putative water transporting proteins in frog urinary bladder by antibody labeling.[266] Polyclonal antibodies against apical membrane proteins from vasopressin-stimulated frog urinary bladder were raised in rabbit.[225] The unfractionated immune serum was found to inhibit osmotic water permeability in intact frog bladders. The serum was purified by immunoadsorption with unstimulated bladders and used to stain apical membrane proteins from stimulated bladder. The purified serum recognized several proteins of sizes 17, 26, 35 and 55 kDa. The affinity purified antibodies inhibited water permeability slightly in intact bladders. In label freeze-fracture electron microscopy using the purified antibodies, gold particles were detected in regions of aggrephore fusion and surface microvilli.[34] These studies suggested that the labeled proteins may be components of the vasopressin-induced water channel in frog bladder.

Recently, Frigeri, Bourguet and coworkers raised polyclonal antibodies against purified apical membranes of frog bladders adapted to high salinity.[68] One of the antisera recognized proteins of 59 and 66 kDa, and stained selectively the apical membrane of some granular cell in frog bladder as well as the apical membrane of principal cells in rabbit kidney. The antibody also stained apical plasma membrane of collecting duct principal cells from Brattleboro rat after stimulation by vasopressin; little staining was observed in tubules from unstimulated rats. It was proposed that the antibody labeled components involved in the vasopressin hydroosmotic response.

Determination of Water Transporter Size by Target Analysis

Radiation inactivation is a unique method to determine protein size in native membranes without protein identification or purification.[128] The principle underlying the radiation activation method is that the probability of irreversibly inactivating a protein by radiation is proportional to protein size. Samples are irradiated by a particle accelerator at low temperature (generally -135°C) to prevent the formation of diffusible free radicals which would invalidate the proportionality between inactivation probability and protein size. The "target size" of a protein is determined from the slope of a ln [activity] vs radiation dose plot. The interpretation of target size values in terms of multi-subunit proteins has been modeled mathematically[260,261] and is discussed further with respect to CHIP28 oligomeric assembly in Chapter 6.

Early studies indicated that the target size for water transport in erythrocytes and kidney proximal tubule apical membrane vesicles was very small, <20-30 kDa.[54,254] Technical limitations restricted the analysis to relatively low-radiation doses where significant changes in activity could be observed only for larger proteins. Recent radiation inactivation studies by Van Hoek, Van Os and coworkers using higher radiation doses showed dose-dependent inactivation of water transport activity (Fig. 4.3). An ~30 kDa target size for water transport was determined in erythrocytes and rat kidney cortical vesicles.[235,236] The 30 kDa size could represent the molecular size of a single water transporting protein, the total size of a multi-subunit water transporting protein, or the size of a functionally independent monomer in a multi-subunit protein. Because lipids and carbohydrates are not inactivated by radiation, the 30 kDa target size provided evidence that the water transporter is at least in part composed of protein. In addition, the search for water channels could be narrowed to proteins of a particular size.

EXPRESSION OF WATER CHANNELS IN XENOPUS OOCYTES

The functional expression of mRNA encoding water channels in *Xenopus* oocytes (Fig. 4.4) provided strong evidence that the water channel is a protein suitable for identification by molecular methods.[294,298] mRNA from a series of tissues was prepared by homogenization in guanidinium thiocyanate followed by standard phenol-chloroform extraction, CsCl gradient centrifugation, and oligo(dT) affinity chromatography. The mRNA was dissolved in water at a concentration of 1 mg/ml and 50 ml of water (control) or mRNA was microinjected into mature, defolliculated oocytes from adult *Xenopus laevis*. After incubation for 48-72 hours for protein translation and targeting, osmotic water permeability was determined by a swelling assay as described in Chapter 2. Measurements were carried out at 10°C to decrease the strongly temperature-dependent endogenous water permeability in oocytes.

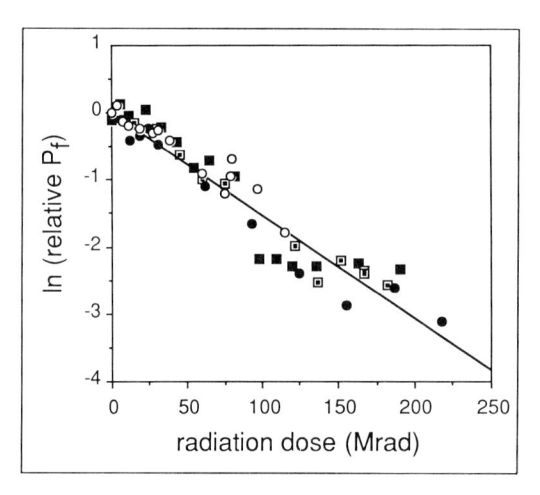

Fig. 4.3. Radiation inactivation analysis of osmotic water permeability in apical membrane vesicles from kidney cortex. P_f was measured in vesicles by stopped-flow light scattering after irradiation with the indicated doses of ionizing radiation. Different symbols correspond to data from different mammalian species. The slope of the fitted lines gives a target size of 30 kDa. See text for explanations. (Adapted from ref. 235.)

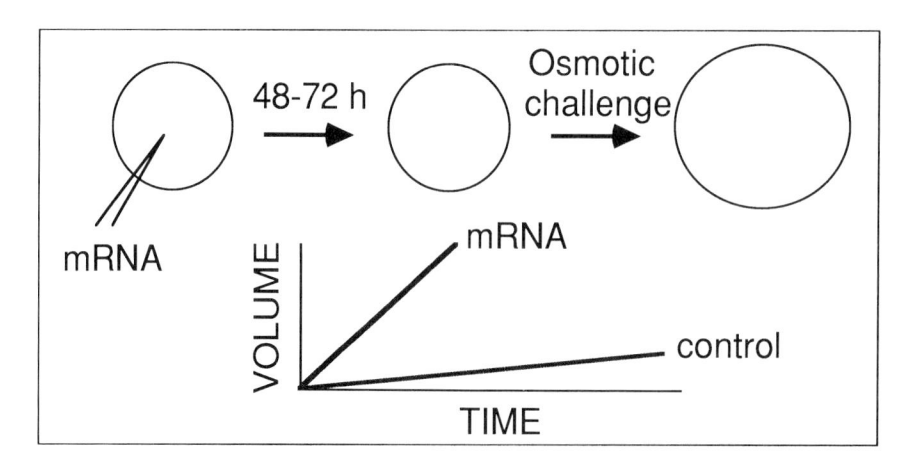

Fig. 4.4. Schematic for water channel expression in Xenopus oocytes. Oocytes were microinjected with tissue mRNA or transcribed cRNA, incubated for 48-72 hours for protein translation and targeting and assayed by measurement of swelling in response to an osmotic gradient.

Figure 4.5 shows representative data for the time course of oocyte swelling in response to a 20-fold dilution of the extracellular Barth's buffer with distilled water. Water permeability was increased significantly by injection of mRNA from kidney and reticulocyte, but not from brain, muscle and liver (see below). Water permeability was maximum at 72 hours after microinjection. P_f in water injected oocytes was 3.6×10^{-4} cm/s. P_f in oocytes injected with 50 ng of unfractionated mRNA was (in 10^{-4} cm/s): 4.0 (rabbit brain), 4.2 (rabbit muscle), 18 (rabbit reticulocyte), 16 (rat renal papilla) and 13 (rat renal cortex).[294] Figure 4.6 summarizes P_f values for oocytes expressing mRNAs from a series of tissues and cultured cells. The increased oocyte water permeability could in principle result from expression of functional water channels encoded by the microinjected mRNA, or the secondary organization of endogenous oocyte membrane components to form a transmembrane conduit for water.

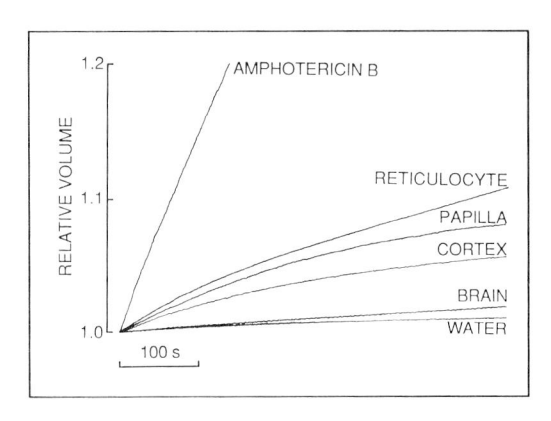

Fig. 4.5. Expression of water channels in Xenopus oocytes. Oocytes were microinjected with water or 5 ng of mRNA isolated from rabbit reticulocyte, rat renal cortex, rat renal papilla and rat brain. Where indicated, the pore-forming agent amphotericin B was added to noninjected oocytes prior to measurement. Oocyte swelling was measured at 10°C after a 20-fold dilution of the extracellular buffer with distilled water. (Adapted from ref. 294.)

To determine whether the expressed water channels had the same characteristics as in the original tissues, inhibition properties and E_a were determined. P_f in oocytes expressing mRNA from rabbit reticulocyte and rat kidney was inhibited to nearly control levels by 0.3 mM $HgCl_2$; the inhibition was reversed by addition of 5 mM mercaptoethanol, indicating that inhibition was not a nonspecific toxic effect of $HgCl_2$.[294] In oocytes expressing mRNA from rabbit reticulocyte, P_f was also inhibited by pCMBS with a concentration-dependence and time course similar to that measured in native rabbit erythrocytes.[222] Oocyte P_f was also increased by expression of mRNA from toad urinary bladder,[298] but not from mRNA isolated from a series of cultured cells including renal epithelial cells (MDCK, LLC-PK1, A6).[297] E_a for expressed kidney and erythrocyte water channels was <4 kcal/mol. In recent expression studies of mRNA from rat kidney cortex and papilla, a small additional increase in oocyte P_f was detected after prolonged exposure to high concentrations of cAMP agonists only in mRNA from renal papilla.[58] Because oocytes probably do not contain a vesicular trafficking mechanism for vasopressin-sensitive water channels as described in Chapter 3, the increased P_f could represent direct phosphorylation of an expressed water channel. Under conditions in which other cAMP-dependent expressed transporters (CFTR[105] and urea transporter[107]) are strongly activated, (e.g., forskolin or Sp-cAMPs for 5-10 min), there was no further increase in water permeability of oocytes expressing mRNA from kidney papilla or cRNA encoding a collecting duct water channel.[148] (See Chapter 7.) There may be other effects of prolonged cAMP stimulation in oocytes.

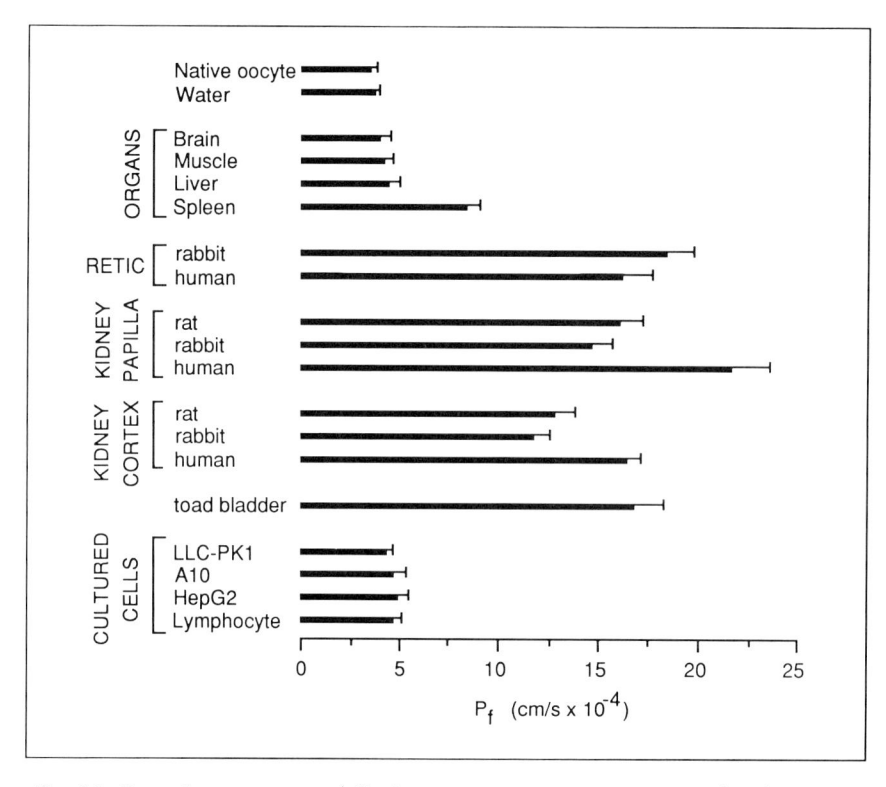

Fig. 4.6. Osmotic water permeability in Xenopus oocytes microinjected with 50 ng of mRNA from a series of tissues and cultured cells. P_f was measured in defolliculated oocytes at 10°C at 48-72 hours after microinjection. "Native oocyte" refers to noninjected defolliculated oocytes and "water" refers to water-injected (control) oocytes.

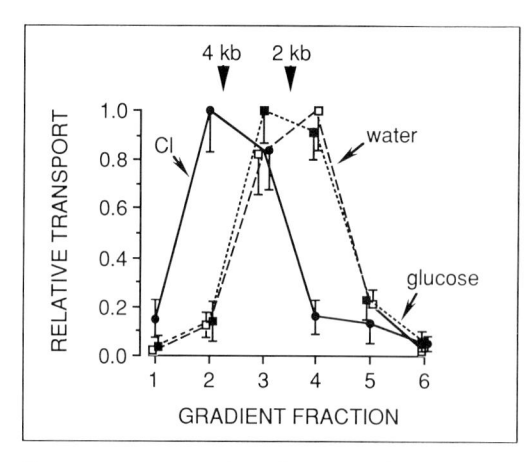

Fig. 4.7. Expression of size-fractionated mRNA from rabbit reticulocytes in Xenopus oocytes. Oocytes were microinjected with ~5 ng of size-fractionated mRNA obtained by sucrose gradient centrifugation. Relative transport of water, glucose and chloride is shown after subtraction of background signal from water-injected oocytes. (Adapted from ref. 292)

Size fractionation studies were carried out with mRNA from rabbit reticulocyte by sucrose density centrifugation in the presence of 10 mM methylmercuric hydroxide (Fig. 4.7).[292] Oocyte water permeability was increased strongly by mRNA in the 2-3 kb size fraction. In parallel measurements of anion and glucose transporters, highest expression was observed at ~4 kb and 2 kb, respectively, in agreement with the known sizes of mRNA encoding the erythrocyte anion exchanger AE1 and glucose transporter GLUT1. These results suggested that the erythrocyte water channel was encoded by one or more mRNAs of 2-3 kb size.

Based on similar data for size fractionated kidney mRNA[295], identification of mRNA encoding water channels from kidney papilla was attempted by expression cloning.[293] Size-fractionated mRNA from rat renal papilla was used to construct a cDNA library; transcribed cRNAs from pooled bacterial colonies were expressed in *Xenopus* oocytes until single colonies were obtained. Three positive colonies in which the expressed cRNA gave increased oocyte water permeability were obtained after screening 25,000 initial colonies. The clone giving highest water permeability had an open reading frame of 2.1 kb encoding a 73 kDa protein. However, subsequent studies showed that the increased oocyte P_f was not inhibited by $HgCl_2$ and that the mRNA and protein had low abundance in kidney. Therefore the cloned proteins had water channel activity, yet were probably not physiologically important water transporters. An unanticipated difficulty with the expression cloning strategy is that many proteins, such as ion channels and glucose transporters, are able to transport water (see below); because of the biophysical constraints limiting the amount of water that can be passed by single water channels, physiologically important water channels must be present in membranes at high density (>10^3 channels/μm^2). (See Chapter 5.)

THE ROLE OF GLUCOSE, ANION AND OTHER TRANSPORTERS IN WATER PERMEABILITY

Prior to identification of specific water transporters, it was believed that water moved through integral membrane proteins that were present in abundant quantities in many cell types. Two candidates were the anion exchanger[32,145,213] and glucose transporter.[64] Fischbarg and coworkers were the first to report increased water permeability in oocytes expressing glucose transporters.[65] As described below, recent data indicates that the anion exchanger band 3 does not transport water, whereas the sodium-independent glucose transporter does transport some water.[292] However, water movement through glucose transporters does not contribute significantly to water permeability in biological membranes. Recently, it was found that the CFTR (cystic fibrosis transmembrane regulator) chloride channel can transport water when it is activated by cAMP agonists.[105]

Oocyte expression studies were carried out to investigate whether the erythrocyte band 3 anion exchanger (AE1) is necessary and/or sufficient to account for the high P_f in erythrocytes. Water and chloride transporter were measured in oocytes expressing transcribed cRNA encoding AE1.[292] Whereas stilbene-inhibited chloride transport increased >5-fold by expression of band 3 protein, there was no increase in oocyte P_f (Fig. 4.8). To determine whether AE1 was necessary for the increase in P_f in oocytes expressing heterologous reticulocyte mRNA, oocytes were coinjected with reticulocyte mRNA and an AE1-antisense DNA. The AE1-antisense DNA suppressed completely the expression of AE1 as demonstrated by the lack of increase in oocyte chloride transport. However, the AE1-antisense DNA did not affect the increase in oocyte P_f, indicating that other distinct mRNAs were responsible for the increased water permeability. Further, the increase in P_f in oocytes expressing reticulocyte mRNA was not inhibited by stilbenes. Taken together, these results indicated that band 3 does not transport water, nor is it required for water transport.

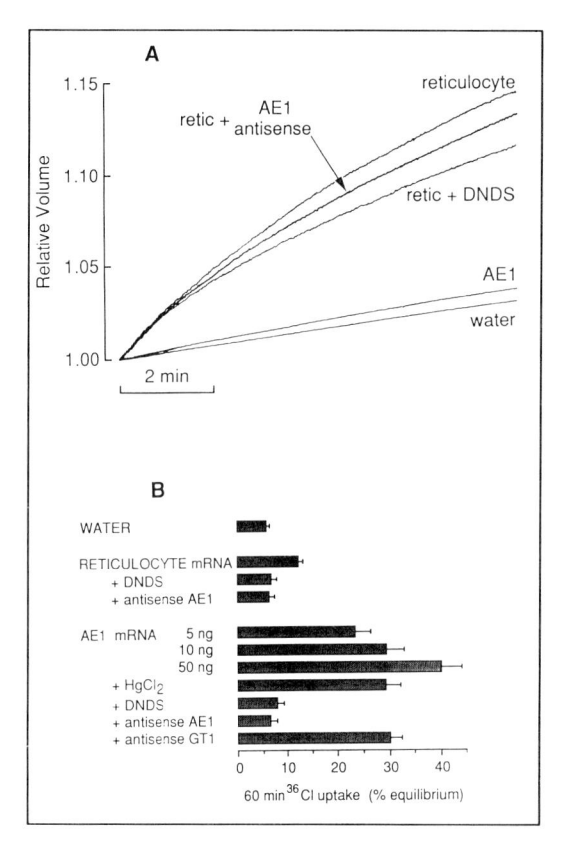

Fig. 4.8. Evidence that the band 3 anion exchange protein (AE1) is not involved in erythrocyte water transport. A. Osmotic water permeability in Xenopus oocytes expressing heterologous reticulocyte mRNA or transcribed cRNA encoding AE1. Oocyte water permeability was not increased by expression of AE1 (compare to water-injected control). Oocyte water permeability was increased by expression of reticulocyte mRNA; however, there was no significant effect of coinjection with antisense AE1 DNA or addition of the anion exchange inhibitor DNDS at the time of the assay. B. Anion exchange in oocytes measured by ^{36}Cl uptake. The increased ^{36}Cl uptake in oocytes expressing reticulocyte mRNA was blocked by DNDS or coinjected with AE1 antisense DNA. ^{36}Cl uptake was increased strongly in oocytes expressing AE1. (From ref. 292)

Similar antisense depletion studies were carried out for the sodium-independent glucose transporter (GLUT1).[292] Glucose transport in oocytes expressing reticulocyte mRNA was increased ~4-fold compared with water-injected controls; the increase in glucose transport was inhibited by cytochalasin B. The increase in water transport was inhibited by $HgCl_2$ but not by cytochalasin B. In oocytes coinjected with reticulocyte mRNA and GLUT1-antisense DNA, glucose transporter expression was suppressed, whereas there was no affect on water transport. These results indicate that the glucose transporter is not required for expression of the erythrocyte water transporter.

Interestingly, it was found that expression of GLUT1 at high levels in *Xenopus* oocytes (>15-fold increase in oocyte glucose transport) was associated with increased water permeability and that the increased water permeability was inhibited by the glucose transport inhibitors cytochalasin B and phloretin.[65,292] However, the contribution of GLUT1 proteins to cell membrane water transport was calculated to be very low. Based on glucose transporter density in erythrocytes, ~0.01% of erythrocyte water permeability could be attributed to movement through GLUT1.[292] In J774 macrophages, Fischbarg and coworkers reported that phloretin and cytochalasin B inhibited water permeability.[64] Water permeability was measured by a laser light scattering method. In similar studies reported subsequently, no effect of glucose transporters could be found by several independent measurement methods including light scattering and confocal microscopy of adherent cells, and stopped-flow light scattering measurements of suspended cells.[59] Although the reasons for the differences in results were not clear, it was proposed that the earlier study was potentially complicated by unstirred layer effects associated with slow solution exchange. Recent studies of reconstituted GLUT1 protein in proteoliposomes[286] and glucose transport inhibitors in renal and intestinal vesicles[49] supported the conclusion that water movement through glucose transporters is not of quantitative physiological significance.

It is generally believed that ion channels are permeable to water when they are open. This hypothesis is supported by mathematical modeling[40,43] and demonstration of streaming potentials in electrophysiological measurements.[1] To test directly whether water can move through an ion channel, P_f was measured in oocytes expressing CFTR chloride channels. It was found that cAMP agonists increased P_f by 4×10^{-4} cm/s at 10°C in oocytes expressing CFTR; there was no effect of cAMP agonists in water-injected oocytes or oocytes expressing the homologous protein MDR1 (human multidrug resistance protein).[105] The increase in water permeability was inhibited by a chloride channel inhibitor [NPPB, 5-nitro-2-(3-phenylpropylamino) benzoate] and was sensitive to oocyte anion composition. In addition, there was a small cAMP-dependent increase in urea permeability in oocytes expressing CFTR but not in control oocytes. The water permeability of CFTR suggested that open ion channels contain a continuous aqueous pathway. The ability of other ion channels to transport water has not been studied; water permeability measurement requires the expression of relatively high densities of channels that remain open after activation. It should be noted that the water permeability associated with open ion channels is probably not of physiological importance because of the low densities of ion channels in cell membranes.

IDENTIFICATION OF THE CHIP28 ERYTHROCYTE WATER TRANSPORTER

During the time in which the oocyte expression assay for water channels was developed and a 30 kDa target size for the putative water channel was reported, Agre and coworkers isolated and cloned a 28 kDa integral membrane protein (CHIP28) from erythrocytes.[52,184] The protein was thought initially to be related to erythrocyte Rh factors, however similarities between CHIP28 and members of the MIP26 family of proteins suggested that CHIP28 may be an erythrocyte water channel. Biochemical and molecular evidence that CHIP28 is a water channel is presented in the next chapter and studies of CHIP28 structure are presented in Chapter 6.

CHAPTER 5

EXPRESSION AND FUNCTION OF THE CHIP28 WATER CHANNEL

CHIP28 is a small hydrophobic protein that is readily isolated from native erythrocytes and mammalian kidney. The cDNAs encoding human and rat CHIP28 have open reading frames of 807 bp and mRNA sizes of ~2.8 kb on Northern blot. Reconstitution of purified CHIP28 in proteoliposomes and expression of CHIP28 in *Xenopus* oocytes and CHO cells indicate that CHIP28 is a selective water transporting protein that excludes ions and small solutes. Residue cysteine 189 in CHIP28 is the site of action of mercurial water transport inhibitors. Residue asparagine 42 is the site of N-linked glycosylation; however, glycosylation is not necessary for CHIP28 water transport. This chapter reviews the functional analysis of reconstituted and expressed CHIP28 water channels.

PURIFICATION AND RECONSTITUTION OF CHIP28 PROTEIN

CHIP28 is an abundant protein with unique properties, making purification of very large (>>10 mg) quantities relatively easy. This makes studies of CHIP28 function in reconstituted proteoliposomes and CHIP28 structure by biochemical and crystallographic methods uniquely attractive.

Figure 5.1 shows a general scheme for purification of CHIP28. It was found that the anionic detergent N-lauroylsarcosine is able to strip virtually all erythrocyte integral membrane proteins other than CHIP28.[52,237] In the initial purification studies, KI-stripped inside-out ghost membranes were prepared from human erythrocytes, and non-CHIP28 integral proteins were extracted by 2% N-lauroylsarcosine. Alternatively, erythrocyte ghost membranes can be stripped directly in up to 6% N-lauroylsarcosine, yielding a right-side out preparation showing predominantly two stained bands on SDS-PAGE (Fig. 5.2, left), a sharp band at 28 kDa representing nonglycosylated CHIP28, and a broad band at 35-60 kDa representing glycosylated CHIP28. CHIP28 identity was confirmed by N-terminus sequence analysis.[237] Enzymatic deglycosylation of CHIP28 by PNGase gives a single band at 28 kDa (Fig. 5.2, right). Osmotic water permeability measured in the stripped membranes containing CHIP28 was similar to that in ghost membranes, providing direct evidence that CHIP28 is the erythrocyte water channel.[237]

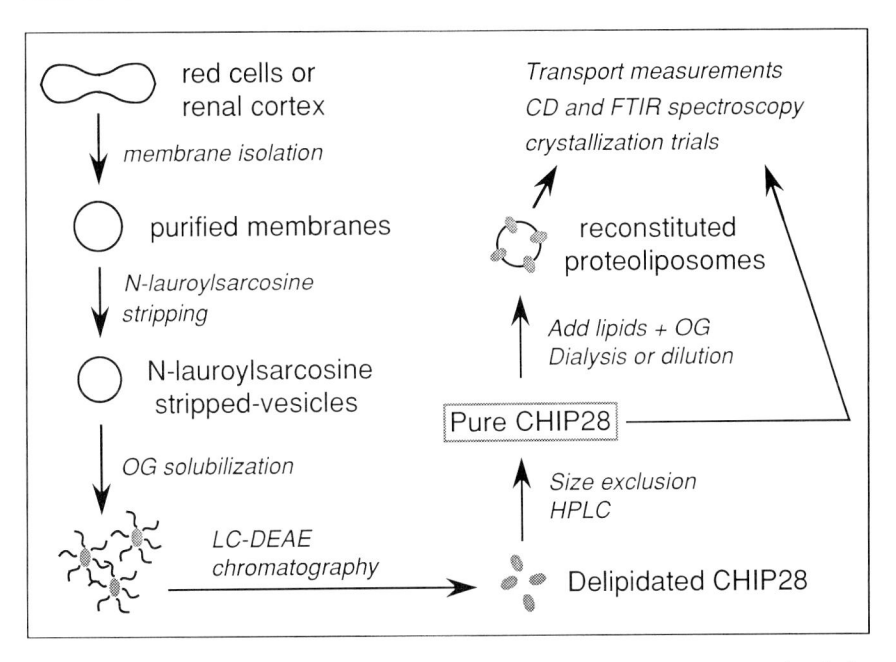

Fig. 5.1. Scheme for purification of CHIP28 protein. Cell membranes are stripped with the anionic detergent N-lauroylsarcosine, detergent solubilized, and subjected to anion exchange (LC-DEAE) and size-exclusion chromatography. See text for details.

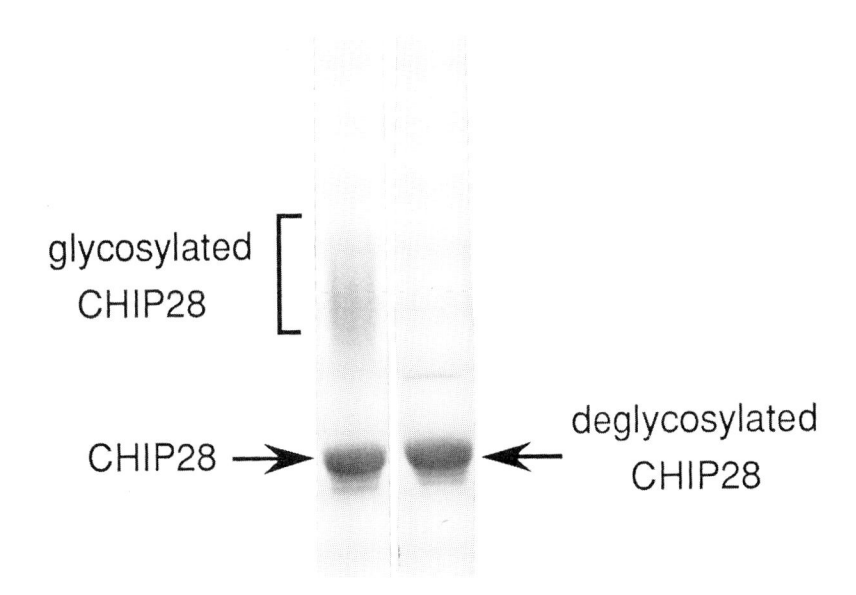

Fig. 5.2. 10% Lamelli SDS-PAGE of purified CHIP28 protein stained with Coomassie blue. Left: HPLC purified CHIP28 showing a nonglycosylated band at 28 kDa and a broad glycosylated band centered at 55 kDa. Right: Enzymatically deglycosylated CHIP28 showing a band at 28 kDa and a thin band for the PNGase enzyme. The gel was heavily loaded to demonstrate purity.

To obtain highly purified CHIP28 protein, CHIP28 in detergent micelles was prepared by suspending the stripped membranes (0.5 mg/ml protein) in EDTA-phosphate buffer containing 200 mM octylglucoside (OG) for one hour.[238] The solubilized membranes were filtered and applied to a DEAE-Sephacel column equilibrated in EDTA-phosphate buffer containing 35 mM OG. Anion-exchange purified CHIP28 was eluted with EDTA-phosphate buffer containing 0.6 M NaCl. The DEAE eluate was filtered through a 0.2 μm filter, concentrated using high-pressure ultrafiltration, and applied to a size-exclusion column (TSK G3000SW). The column was eluted with 20 mM Na phosphate, 100 mM NaCl, 35 mM OG (pH 7.0) and the eluate was monitored at 280 nm. CHIP28 eluted as an apparent dimer with a single peak containing both the nonglycosylated and glycosylated forms. These two forms could not be separated by lectin and phenylborinate[25] chromatography under a variety of conditions, suggesting strong noncovalent association.

Several methods have been applied to reconstitute CHIP28 water channels, including Triton X-100 solubilization, lipid addition and Triton removal by Biobeads[237], and OG solubilization and detergent dilution.[238,287] In our recent studies, lipid vesicles containing PC, PI and cholesterol (molar ratio 11:1:11) were prepared by reverse phase evaporation.[238] An ether solution was hydrated with 3.2 ml EDTA-phosphate buffer, vortexed vigorously and evaporated with N_2. The milky suspension was dispersed by homogenization in 10 ml of EDTA-phosphate buffer to yield lipid vesicles, and then diluted to 100 ml and centrifuged. To reconstitute CHIP28, lipid vesicles were dissolved at 40°C in EDTA-phosphate buffer containing 150 mM OG, cooled to 20°C, and mixed with purified OG-solubilized protein. The suspension (lipid:protein, 10 g/g) was either dialyzed against detergent-free EDTA-phosphate buffer or slowly diluted with the buffer to decrease the concentration of OG to 5 mM and centrifuged. The pellet contained reconstituted proteoliposomes.

Figure 5.3 (panel A) shows the time course of osmotic water transport in N-lauroylsarcosine-stripped erythrocyte membranes, reconstituted proteoliposomes and (protein-free) liposomes. The high-water permeability in the stripped vesicles and proteoliposomes was inhibited by >98% by 0.1 mM $HgCl_2$. The single channel water permeability (p_f), calculated from P_f and CHIP28 protein concentration, was ~10^{-13} cm^3/s at 37°C. This value assumes that the CHIP28 monomer is functional. The p_f of 10^{-13} cm^3/s, taken together with erythrocyte geometry and measured P_f of 0.02 cm/s, indicates that erythrocytes contain ~2 x 10^5 functional CHIP28 monomers. The high density of CHIP28 predicted by this calculation agrees well with quantitative immunoblots showing that ~3% of total erythrocyte protein is CHIP28.[185] CHIP28 is thus necessary and sufficient to account for the high erythrocyte water permeability. Similar purification and reconstitution studies have been carried out for CHIP28 purified from mammalian kidney proximal tubule.[295]

The reconstituted proteoliposomes containing functional CHIP28 water channels were used to determine whether CHIP28 also transports small solutes and protons. Stopped-flow light scattering measurements showed that water transport was increased >50-fold by incorporation of CHIP28, while urea transport was not changed (Fig. 5.3, panel B).[237,287] Similarly, there was no difference in passive proton permeability in the CHIP28-containing and protein-free liposomes. These results indicate that CHIP28 is a selective water channel that excludes protons and small solutes. Biologically, the water transport selectivity of CHIP28 makes good sense because of the high density of water channels in water-permeable membranes; cell permeabilities to ions and solutes would be outrageously high if CHIP28 were not selective. Estimation of the CHIP28 aqueous pore size based on p_f values and equations describing flow through channels gives a diameter of ~0.2 nm,[295] consistent with the exclusion of urea and monovalent ions.

EXPRESSION OF CHIP28 IN XENOPUS OOCYTES

Human CHIP28 was cloned from a human fetal liver cDNA library using degenerate oligonucleotides derived from N-terminus sequence information.[184] Rat kidney CHIP28 (also referred to as CHIP28k) was cloned by homology from a rat kidney cDNA library using as a probe the full-length coding region of CHIP28k.[295] CHIP28k contains an 807 bp open reading frame encoding a 28.8 kDa protein having 87% nucleotide identity and 94% amino identity to human CHIP28. Total mRNA size was ~2.8 kb by Northern blot of rat kidney RNA. A recent study localized the human

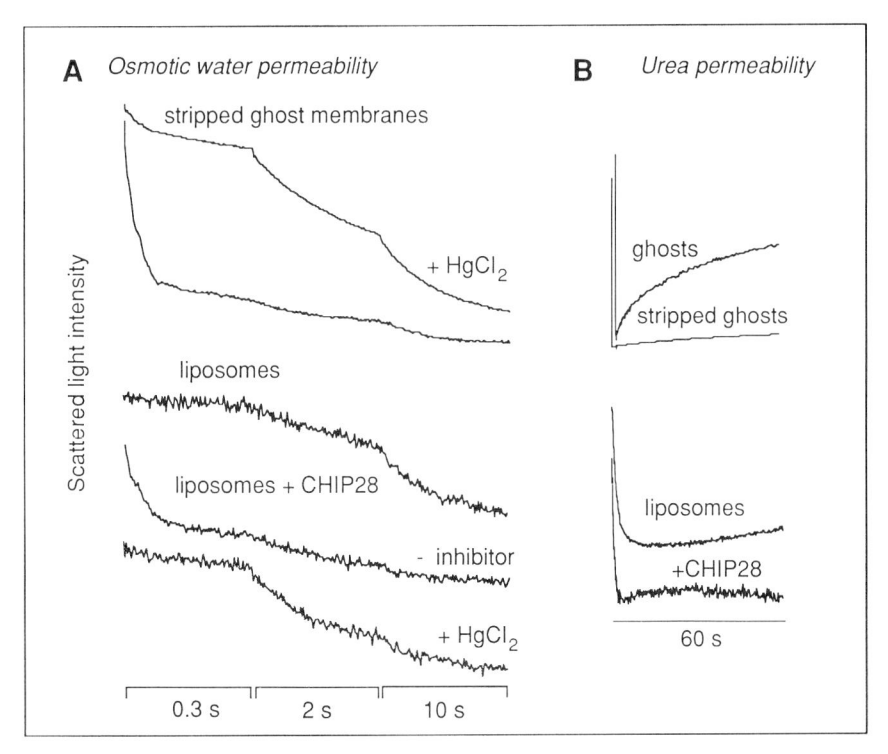

Fig. 5.3. Osmotic water and urea permeability in erythrocyte membranes and liposomes reconstituted with purified CHIP28. A. Stopped-flow light scattering measurement of osmotic water permeability at 10°C in N-lauroylsarcosine-stripped erythrocyte membranes, (protein-free) liposomes, and proteoliposomes reconstituted with CHIP28. Where indicated, 0.3 mM HgCl₂ was present. B. Urea permeability measured from the osmotic response to an inwardly-directed urea gradient. Urea permeability was non-zero in ghosts but very low in N-lauroylsarcosine-stripped erythrocyte membranes and liposomes.

CHIP28 gene to chromosome 7p14-15[47] which is near the CFTR gene locus. Figure 5.4 shows an amino acid sequence alignment for rat and human CHIP28, together with the sequence of MIP26 (Major Intrinsic Protein of lens of 26 kDa). Additional alignment with other MIP26 family members is given in Chapter 7. Two highly conserved sequences called "NPA boxes" are seen, as well as a number of other highly conserved regions. A more complete discussion of "water channel family relations" is provided in Chapter 7.

A Kyte-Doolittle hydropathy plot of rat CHIP28 is shown in Figure 5.5. CHIP28 is an hydrophobic protein with short polar segments corresponding to putative extramembrane domains, and a longer C-terminus polar segment. The hydropathy analysis is consistent with up to seven membrane-spanning domains. Based on biochemical evidence suggesting that the N- and C-termini are cytoplasmically oriented, it was suggested that CHIP28 spans the membrane six times,[184] and that the first and second halves of CHIP28 are "tandem repeats" having similar transmembrane topology. CHIP28 has two potential sites for N-linked glycosylation (Asn 42 and Asn 205), four cysteine (C) residues, four tryptophan (W) residues and four potential phosphorylation

```
MIP26:     1  MWELRSASFWRAICAEFFASLFYVFFGLGASLR    WAPGPLHVLQVALAFGLALATLVQAVGHISGAHVNPAVTFAFLVG
CHIP28:    1 MASEFKKKLFWRAVVAEFLATTLFVFISIGSALGFKYPVGNNQTAVQDNVKVSLAFGLSIATLAQSVGHISGAHLNPAVTLGLLLS
CHIP28k:   1 MASEFKKKLFWRAVVAEFLAMTLFVFISIGSALGFNYPLERNQTLVQDNVKVSLAFGLSIATLAQSVGHISGAHLNPAVTLGLLLS

MIP26:    79 SQMSLLRAICYMVAQLLGAVAGAAVLYSVTPPAVRGNLALNT
CHIP28:   87 CQISIFRALMYIIAQCVGAIVATAILSGITSSLTGSNLGRND
CHIP28k:  87 CQISILRAVMYIIAQCVGAIVASAILSGITSSLLENSLGRND

MIP26:   121 LHPGVSVGQATIVEIFLTLQFVLCIFATYDERRNGRLGSVALAVGFSLTLGHLFGMYYTGAGMNPARSFAPAILT
CHIP28:  129 LADGVNSGQGLGIEIIGTLQLVLCVLATTDRRRRDLGGSAPLAIGLSVALGHLLAIDYTGCGINPARSFGSAVIT
CHIP28k: 129 LARGVNSGQGLGIEIIGTLQLVLCVLATTDRRRRDLGGSAPLAIGLSVALGHLLAIDYTGCGINPARSFGSAVLT

MIP26:   196 RNFTNHWVYWVGPVIGAGLGSLLYDFLLFPRLKSVSERLSILKGSRPSESNGQPEVTGEPVELKTQAL
CHIP28:  204 HNFSNHWIFWVGPFIGGALAVLIYDFILAPRSSDLTDRVKVWTSGQVEEYDLDADDINSRVEMKPK
CHIP28k: 204 RNFSNHWIFWVGPFIGSALAVLIYDFILAPRSSDFTDRMKVWTSGQVEEYDLDADDINSRVEMKPK
```

Fig. 5.4. Amino acid sequence alignment for human CHIP28, rat CHIP28k and bovine MIP26.

sites (arrows). In vitro translation of rat CHIP28 in rabbit reticulocyte lysate supplemented with dog pancreatic microsomes gave bands at ~28 and 32 kDa, corresponding to nonglycosylated and glycosylated CHIP28, respectively.[295] Translation of a CHIP28 fragment (truncated at bp 343 by PstI) gave full glycosylation, suggesting that Asn 42 was glycosylated and Asn 205 was not glycosylated.

The first indication that CHIP28 encoded a water channel was obtained by expression of transcribed cRNA in *Xenopus* oocytes.[185] Oocytes expressing CHIP28 were >10-fold more water permeable than water-injected (control) oocytes. The increased water permeability was inhibited by $HgCl_2$. Similar high-water permeability was measured in oocytes expressing rat kidney CHIP28[295] as shown in Figure 5.6. Water permeability was nearly as high as that obtained when oocytes were incubated with the artificial pore-forming agent amphotericin B. Oocyte water permeability was inhibited by $HgCl_2$ and the inhibition was reversed by mercaptoethanol. Mea-

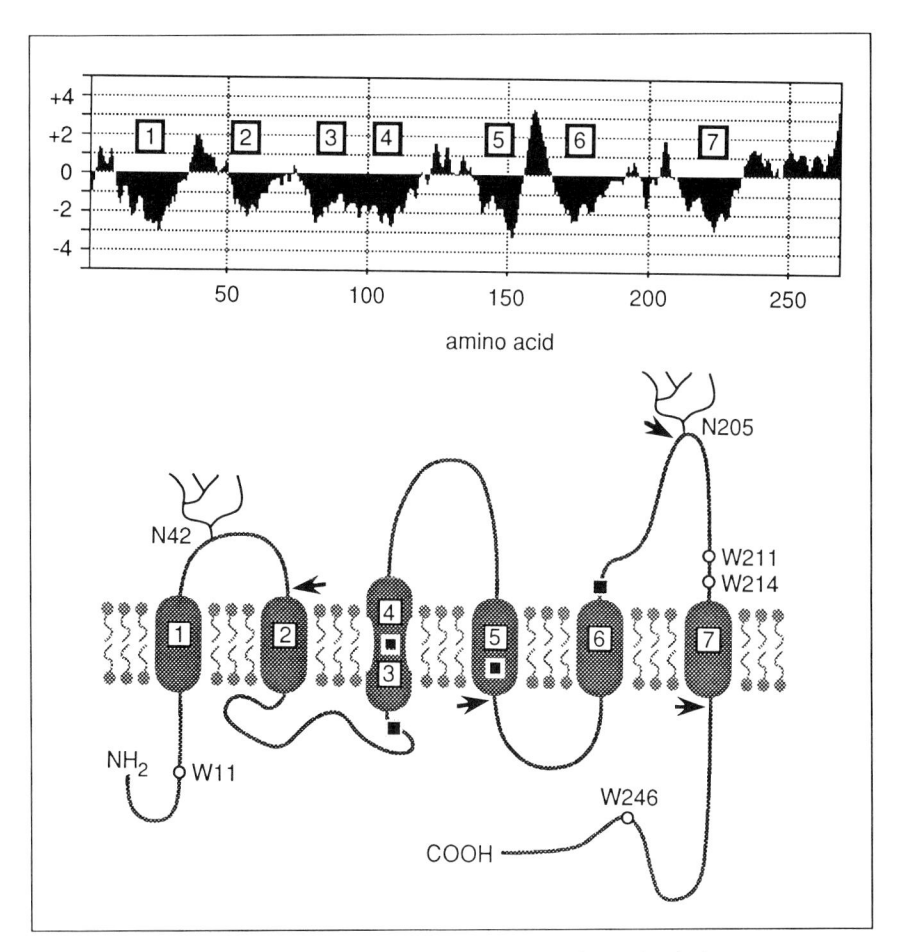

Fig. 5.5. Kyte-Doolittle hydropathy plot and proposed topology of rat kidney CHIP28 water channel. The hydropathy plot shows 7 hydrophobic regions that are potential membrane-spanning domains. The topology proposed initially shows two possible sites for N-linked glycosylation (N42 and N205), four cysteines (boxes), four tryptophans (W11, W211, W214 and W246), and four consensus sequences for phosphorylation (arrows).

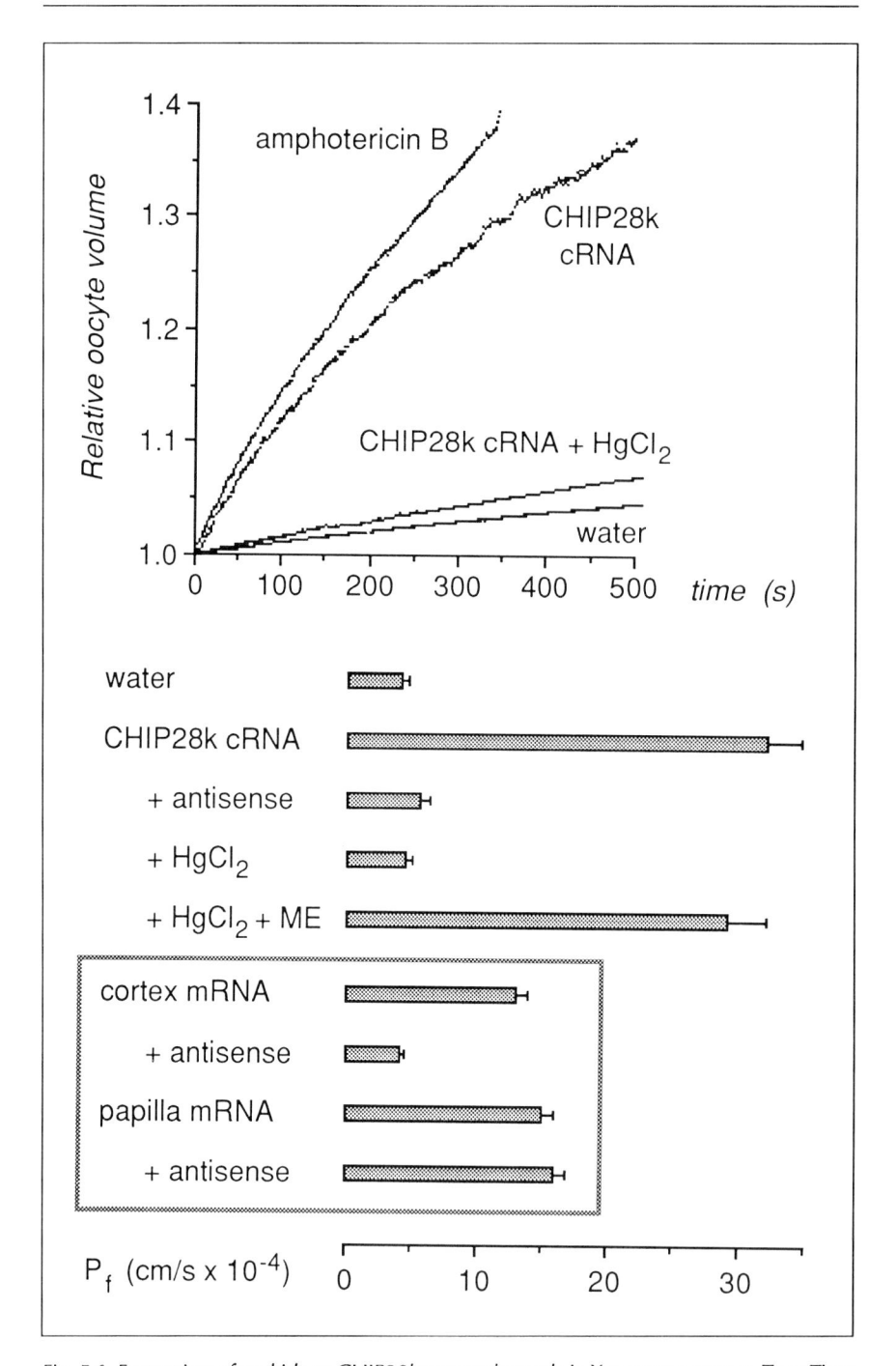

Fig. 5.6. Expression of rat kidney CHIP28k water channels in Xenopus oocytes. Top: Time course of swelling in oocytes injected with water (control) or transcribed cRNA encoding CHIP28k. Where indicated, amphotericin B (0.2 mg/ml) or HgCl₂ (0.3 mM) were present. Bottom: Averaged results (SE) for a series of oocytes. Where indicated, oocytes were coinjected with CHIP28k antisense cRNA. "ME" indicates addition of β-mercaptoethanol after HgCl₂. The data shown in the box represent hybridization-depletion experiments in which antisense CHIP28k cRNA was coinjected with heterologous mRNA isolated from rat kidney cortex or papilla. (Adapted from ref. 295.)

surements of oocyte ion conductance by two-electrode voltage clamp indicated that CHIP28 was not ion permeable,[185,295] consistent with the reconstitution data showing that CHIP28 transported water selectively.

To determine whether water channels different from CHIP28 might be expressed in kidney, oocytes were coinjected with total mRNA from kidney cortex or papilla together with antisense CHIP28 cRNA.[58,295] In cortex, the majority of the increase in oocyte water permeability conferred by expression of native mRNA was blocked by the antisense CHIP28 cRNA (Fig. 5.6, bottom). In contrast, in papilla, little of the increase in oocyte water permeability was blocked by the antisense cRNA. These results suggested that the renal papilla contains additional water channel(s) that do not hybridize well with antisense CHIP28 cRNA. The water channels from renal cortex expressed in oocytes therefore consist of CHIP28 and possibly other homologous proteins; however, the existence of additional water channels in renal cortex that are not expressed well in *Xenopus* oocytes cannot be ruled out.

STABLE EXPRESSION OF CHIP28 IN MAMMALIAN CELLS

The development of a stably transfected cell line for water channels serves a number of purposes, including the analysis of water channel: a) function, b) biogenesis and posttranslational processing, c) intracellular targeting, and d) interaction of other cellular components. In addition, high-expression stable cell lines can be useful for isolation and purification of water channel proteins, including mutant and cloned water channels for which there is no adequate native tissue source. There were several potential difficulties in the development of a cell line that stably expresses functional water channels. Because of the small single channel water permeability of CHIP28 (~10^{-13} cm^3/s) and the relatively high-endogenous water permeability of biological membranes (~10^{-3} cm/s), a high-membrane density of functional water channels (>100 per μm^2) was required in transfected cells to increase water permeability measurably. In addition, it was not known *a priori* whether the increase in water permeability in cell plasma membranes and intracellular vesicles would affect cell viability. Although water movement across membranes is always secondary to osmotic gradients and hydrostatic forces, some cell processes, such as fusion and sorting of vesicles, are associated with volume changes. It has been proposed that the kinetics of volume change may be important for vesicle fusion and other intracellular events.

The CHO-K1 cell line was chosen for stable expression of CHIP28.[147] A Hind3-XbaI fragment from plasmid pSP64.CHIP28k was subcloned into mammalian expression vectors pRc/CMV and pRc/RSV (Invitrogen). CHO-K1 (wild type) cells were transfected with Lipofectin. Cells were plated at a density of 5 x 10^5 cells per 60 mm diameter dish 12 hours before transfection. The cells were washed three times, and 10 mg/ml Lipofectin and 5 μg/ml of recombinant plasmid (or vector only) was added in serum-free medium to the washed CHO-K1 cells. Serum and Geneticin (500 μg/ml) were subsequently added. The cells remaining after 14 days of selection were plated at clonal densities. Clonal cell populations were screened initially by anti-CHIP28 antibody staining of permeabilized cells. Northern and Western blots of transfected cells showed CHIP28k mRNA and protein expression; blots of the mock-transfected cells (vector alone) were negative.

Functional and localization studies were carried out on the stably transfected cells. Osmotic water permeability was measured by stopped-flow light scattering in intact suspended cells (obtained by low Ca/EGTA, no trypsin) and in isolated plasma membranes, Golgi and endoplasmic reticulum (obtained by sucrose gradient centrifugation). Figure 5.7 shows that water permeability was remarkably higher in intact cells (upper left), plasma membrane vesicles, and Golgi (right) in the CHIP28k-expressing than in the control cells. The increased water permeability in the CHIP28k-expressing cells was inhibited by $HgCl_2$. Water permeability was low in endoplasmic reticulum vesicles. Immunoblots with anti-CHIP28 antibody showed strong expression of mature CHIP28k (glycosylated form present) in plasma membranes and Golgi, whereas CHIP28k in endoplasmic reticulum was not glycosylated. These results indicate that CHIP28k attains functional maturity in Golgi. Quantitative Western analysis of the cell line with highest CHIP28 expression indicated ~8 x 10^6 copies of CHIP28 per cell; blots of the purified plasma membrane fraction indicated that ~4% of plasma membrane protein was CHIP28.

In native proximal tubule which expresses CHIP28 protein, functional water channels are found on plasma membranes, clathrin-coated vesicles and endosomes.[193] To determine whether CHIP28k was present in endosomes in CHIP28k-expressing CHO cells, endosomes were labeled with 6-carboxyfluorescein and osmotic water permeability was measured in a microsomal pellet containing the fluorescent endosomes by the stopped-flow fluorescence quenching assay. Water permeability was high in the majority of endosomes in the CHIP28k-expressing cells (Fig. 5.7, lower left). After inhibition by $HgCl_2$, water permeability decreased to values obtained in endosomes from control cells. Immunogold electron microscopy of the CHIP28k-expressing cells (see Chapter 8, Fig. 8.2) showed localization of gold particles on plasma membrane, Golgi and intracellular vesicles. A preimmune serum control was negative.

These findings indicate that the development of stably transfected cell lines expressing functional water channels is possible. The mammalian cell expression system should be useful for expression and analysis of new and mutated water channels.

Mechanism of Inhibition of Water Permeability by Mercurials

The inhibition of cell membrane water permeability by mercurials has been taken as a signature of proteinaceous water channels in a variety of tissues and subcellular fractions, including erythrocytes,[149] kidney proximal tubule,[159,181,234] amphibian urinary bladder[112,113] and various endosomes and intracellular vesicles.[204,257,283] Mercurials also inhibit the water permeability of purified reconstituted CHIP28 protein,[237,238,287,295] and of expressed CHIP28 in *Xenopus* oocytes[185,295] and CHO cells.[147] The mercurial compounds $HgCl_2$ and pCMBS (p-chloromercuribenzenesulfonate) inhibit water permeability rapidly; inhibition is reversed by addition of the sulfhydryl reducing compounds dithiothreitol or β-mercaptoethanol. The chemistry of mercurial binding to proteins and the reversibility by reducing agents indicates an interaction at one or more cysteine residues.

CHIP28 contains four cysteine residues at positions 87, 102, 152 and 189. Site-directed mutagenesis studies in human and rat CHIP28 provided

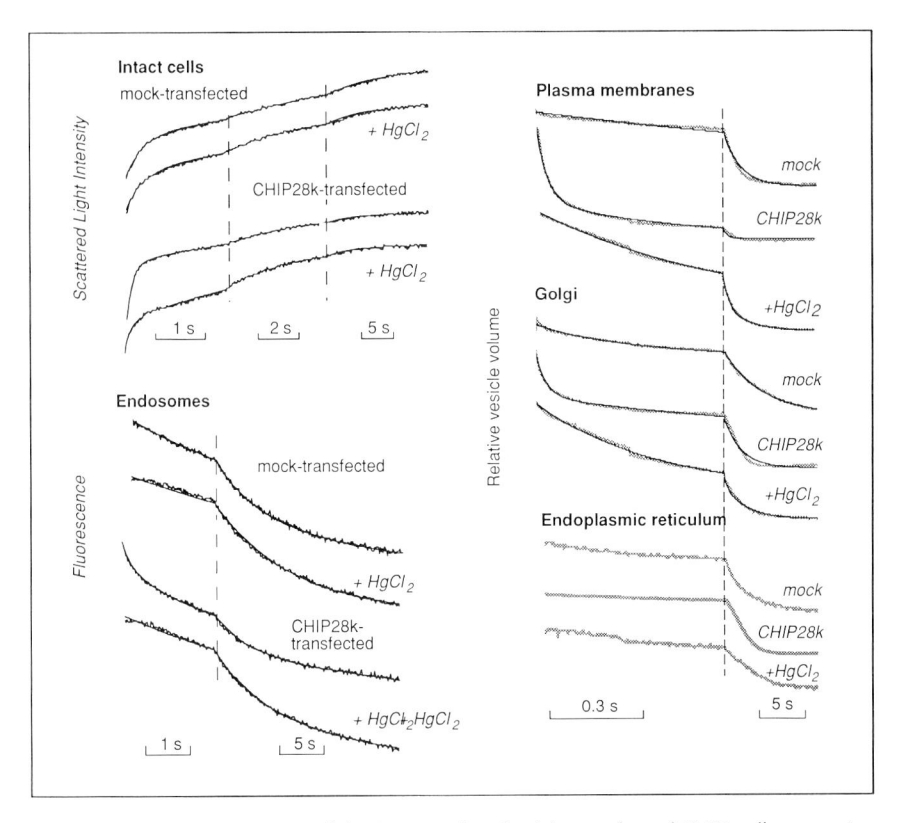

Fig. 5.7. Osmotic water permeability in control and stably-transfected CHO cells expressing rat CHIP28k. Where indicated, 0.3 mM HgCl₂ was added. Upper left: Stopped-flow light scattering measurements in freshly suspended cells. Lower left: Fluorescence self-quenching measurements in carboxyfluorescein-labeled endosomes. Right: Stopped-flow light scattering measurements in purified subcellular fractions. With the exception of endoplasmic reticulum, note the increased water permeability in membranes derived from the CHIP28k-expressing cells. (Adapted from ref. 147.)

direct evidence that cysteine 189 is the major site for mercurial inhibition of water permeability.[186,296] Mutant cDNAs were transcribed and expressed in *Xenopus* oocytes. Mutation of cysteines 87, 102 and 152 to serines, individually or together, had no effect on water permeability or on the inhibition of water permeability by HgCl$_2$ (Fig. 5.8).[296] Mutation of cysteine 189 to serine or glycine had no effect on water permeability, but abolished completely the inhibition by HgCl$_2$. Mutation of cysteine 189 to tryptophan inhibited water permeability by either blocking water channel function directly or water channel targeting to the oocyte plasma membrane.

To determine whether cysteine 189 resides at the external or cytoplasmic face of CHIP28, an impermeable mercurial compound (pCMB-dextran) was synthesized.[296] The pCMB-dextran compound inhibited water permeability effectively in intact erythrocytes but not in inside-side out erythrocyte vesicles (Fig. 5.9), indicating that cysteine 189 faces the external surface. This topology is consistent with studies described in Chapter 6.

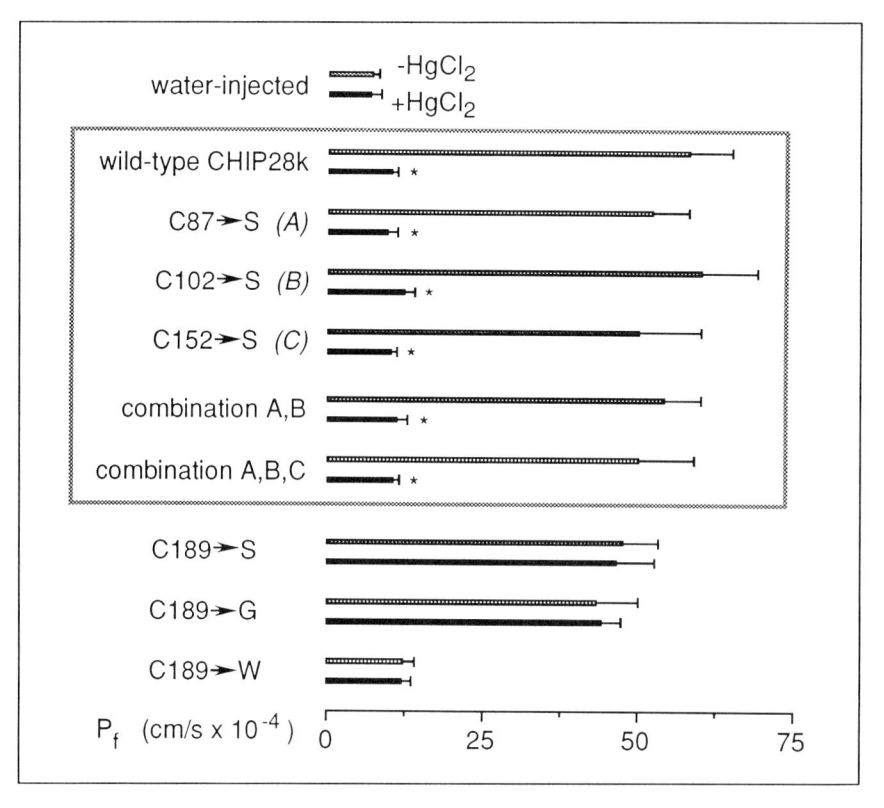

Fig. 5.8. Site-directed mutagenesis of cysteine residues in rat kidney CHIP28. Osmotic water permeability in Xenopus oocytes microinjected with water or transcribed cRNA encoding wild-type or mutant protein. Mutation of cysteines at positions 87, 102 and 152 had no effect on P_f or $HgCl_2$ inhibition. Mutation of cysteine 189 to serine or glycine abolished $HgCl_2$ inhibition, whereas mutation to tryptophan (W) gave nonfunctional protein. (Adapted from ref. 296)

Effects of Glycosylation on CHIP28 Function

Densitometric analysis of Coomassie blue-stained gels of purified CHIP28 suggest that ~50% of CHIP28 monomers are glycosylated. The detailed pattern of glycosylation differs in different organs and species. The oocyte expression studies described above suggest that N-linked glycosylation of CHIP28 occurs only at residue Asn 42; preliminary analysis of tryptic fragments by electrospray mass spectrometry confirms that the potential N-linked glycosylation site Asn 205 is not glycosylated.[239]

Several lines of evidence suggest that CHIP28 glycosylation is not critical for its water transporting function. Water permeability in oocytes expressing mutants N42T and N205T is not different than that in oocytes expressing wild-type CHIP28, nor is there a difference in the inhibition by $HgCl_2$.[296] Enzymatic deglycosylation of CHIP28 in right-side-out vesicles by PNGase resulted in the disappearance of glycosylated CHIP28 from Commassie blue-stained gel and immunoblot, but did not alter water permeability as assayed by stopped-flow light scattering. Reconstitution of deglycosylated CHIP28 in proteoliposomes showed full water transport function. Incubation of CHIP28-expressing CHO cells with tunicamycin for 48 hours decreased CHIP28 glycosylation but did not affect water permeability. These results indicate that N-linked glycosylation of CHIP28 is not essential for water transport of CHIP28 or its membrane targeting; it is not known whether glycosylation affects CHIP28 monomer-monomer interactions and/or oligomeric assembly in membranes.

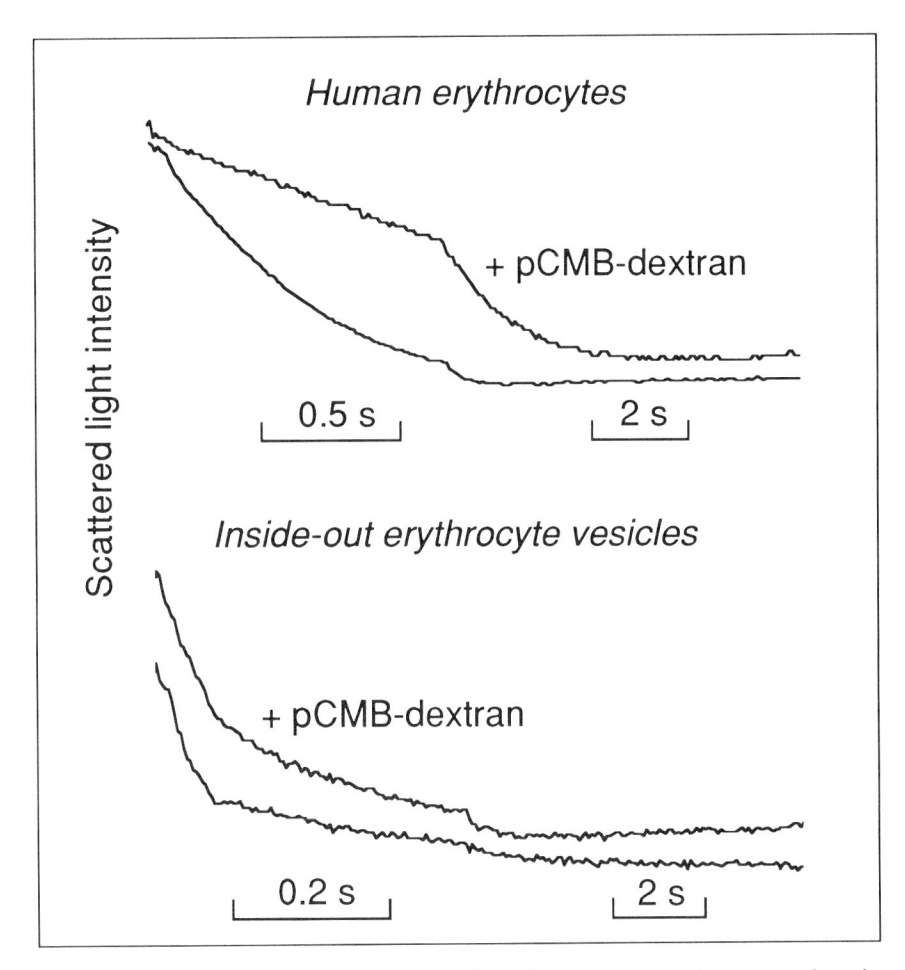

Fig. 5.9. Inhibition of osmotic water permeability in intact human erythrocytes and inside-out erythrocyte vesicles by pCMB-dextran. Water permeability at 10°C was measured by the stopped-flow light scattering technique. Where indicated, pCMB-dextran (50 mg/ml) was incubated with membranes for 30 min prior to measurements. (From ref. 296.)

CHAPTER 6

STRUCTURE AND WATER-TRANSPORTING MECHANISM OF CHIP28

A fundamental question is how CHIP28 transports water. Information in Chapter 5 suggests that CHIP28 forms a narrow aqueous channel that excludes all ions and solutes other than water. The structural data in this chapter suggest that independently functioning CHIP28 monomers are assembled as tetramers in membranes, and that each monomer contains four membrane-spanning α–helical domains forming a central aqueous channel. Validation of this tentative model for CHIP28 structure will require high-resolution diffraction studies on suitable CHIP28 crystals.

HYDROPATHY ANALYSIS OF CHIP28 SECONDARY STRUCTURE

Analysis of the predicted primary sequences of CHIP28 and members of the MIP26 family of proteins provides useful information about putative CHIP28 structure.[238] Figure 6.1 shows a generalized hydropathy analysis of CHIP28 (left) and MIP26 (right). The upper plots are conventional Kyte-Doolittle hydropathy plots with an amino acid window of 7. Based on hydropathy indices alone, segments labeled 1, 2, 4, 5, 6 and 8 can form putative transmembrane regions. The second and third plots in each column are "weighted Fourier-transform hydropathy plots" based on Jähnig's original formulation.[118] Amphiphilic structures of periodicity p were tentatively identified by examination of the "p-weighted" Fourier sum,

$$H_p(n) = \Sigma\, h(n-i)\, (1+\cos[2\pi i/p]) \,/\, \Sigma\, (1+\cos[2\pi/p]) \quad \text{(Eqn. 1)}$$

where $H_p(n)$ is the "p-weighted" hydropathy of residue n, and h(n) is the Kyte-Doolittle hydropathy value of residue n. The sum is taken over the window of residues i=n-w to n+w. Values of p=3.6 and 2 were used to identify putative amphiphilic α–helix and β–sheet, respectively. The hydropathy pattern of an amphiphilic β–sheet will oscillate strongly with frequency (period) p=2. According to Jähnig, the minima and maxima H(n) values should extend beyond the range 0.4-1.6 for p=2; the range would be 0.8-1.6 for p=3.6. The p=2 plots for CHIP28 and MIP26 show that the first part of region 2, region 3, part of region 5 and region 6 show strong oscillation and could contain β–structure. Analysis with a periodicity of 3.6 suggests that regions 4 and 8 might be amphipathic α–helix. β–turns in connecting loops between transmembrane regions were detected by calculation of Chou-Fasman turn propensities. The regions of high-turn propensity generally correspond to interfacial regions between the polar and hydrophobic regions observed on the unweighted hydropathy plots (A and E). Based on the generalized hydropathy analysis, the predicted α–helix content for membrane-spanning segments is 39-47% for CHIP28 and MIP26.[238] This prediction is subject to direct experimental verification below.

It is emphasized that conclusions derived from primary amino acid sequence of hydrophobic membrane proteins must be viewed cautiously. Unlike the substantial base of information available for modeling the structure of soluble proteins, there is little comparable information for integral membrane proteins. For this reason, a series of studies are described below to elucidate some preliminary features of CHIP28 structure and water-transporting mechanism.

Fig. 6.1. Hydropathy and amphipathy plots for CHIP28 (left) and MIP26 (right). A. and E. Kyte-Doolittle hydropathy profiles with periodicity 1, window width 7; B. and F. Plots with periodicity 2 (see Eqn. 1) and window width 11 to evaluate amphiphilic β–sheet. C. and G. Plot with periodicity 3.6 and window width 19 to evaluate amphiphilic α–helix. D. and H. Chou-Fasman turn propensity. The horizontal lines in A-C and E-F represent hydropathy values H(n) of 1.6, 0.8 and 0.4 (upper to lower) and are used to evaluate hydrophobicity and amphipathy. The horizontal lines in D and H represent probability values. Boxes 1-8 are referred to in the text. (From ref. 238.)

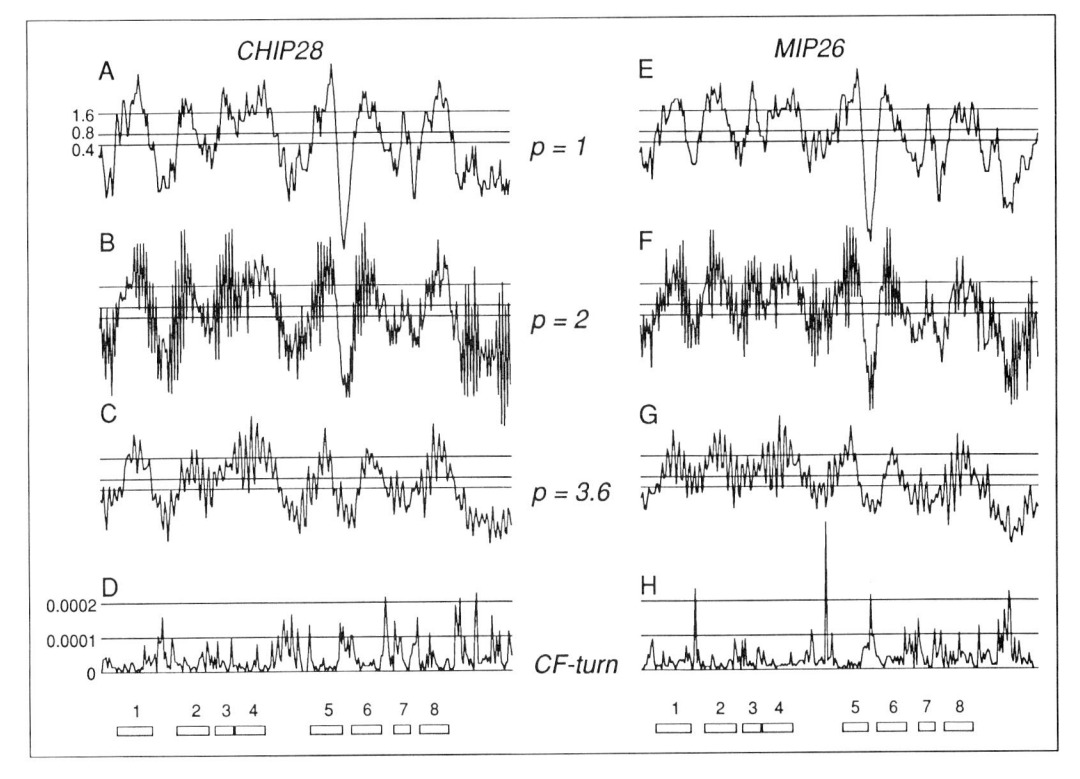

Spectroscopic Analysis of CHIP28 Secondary Structure

Although a multispanning α–helical motif is often assumed for membrane channels and receptors based on analysis of hydropathy, a β–barrel structure (as determined for bacterial porin channels) is also possible. Prior to analysis of CHIP28 membrane topology, studies were carried out to determine whether the secondary structure of CHIP28 had sufficient α–helical content to support a model of multiple membrane-spanning helices. Based on the hydropathy analyses above, CHIP28 should contain at least ~35 % helical content if four or more helical segments spanned the membrane; however, if CHIP28 formed a β–barrel structure, the helical content would be near zero.

To determine the structural motif, circular dichroism (CD) and Fourier transform infrared (FTIR) spectroscopy studies were performed on purified CHIP28 in detergent and membranes.[238] CD is based on the ability of secondary protein structures to differentially transmit left- and right-handed circularly polarized light. Characteristic spectral patterns have been associated with specific structural motifs, e.g., α–helix, β–sheet, β–turn, "unordered structures". CD spectra were obtained with a Jasco J500A spectropolarimeter (JASCO Inc., Tokyo, Japan). CD spectra (with baseline spectra subtracted) were converted to molar ellipticity [[θ] = (θ·MRW)/(10·l·c), deg·cm^2/decimol] using the measured ellipticity (θ, deg), protein concentration (c, g/cm^3), cuvette pathlength (l, 0.05 cm) and mean-residue weight (MRW, 114.3 g/mol). Secondary structure was analyzed by "spectral decomposition" methods in which the contributions of "basis spectra" (representing the spectra corresponding to pure structural forms) to the measured CD spectrum were fitted mathematically. Because of uncertainties in the choice of basis spectra to describe a membrane-associated protein, we used both the four basis spectra set (α–helix, β–sheet, β–turn and "other") of Chang et al[36] and the five basis spectra set ("transmembrane" α–helix, α–helix, β–sheet, β–turn and "other") of Park et al.[173] Data were analyzed both by fixing absolute molar ellipticities (constrained analysis) and allowing ellipticities to float (unconstrained analysis).

Figure 6.2 (panel A) shows a CD spectrum of purified CHIP28 in the detergent β–octylglucoside (OG). Unconstrained analysis using the Chang et al basis spectra (dashed curve) gave 42% α–helix, 43% β–forms (sheet + turn) and 15% unordered structure. Constrained analysis with the Chang et al basis spectra gave 52% α–helix, 31% β–forms and 18% unordered structure; results obtained with the Park et al. basis spectra were in qualitative agreement with these values. Analysis of spectra obtained for reconstituted CHIP28 in proteoliposomes gave 40% α–helix, 40% β–forms (sheet + turn) and 20% unordered structures; addition of the inhibitor HgCl$_2$ did not change CHIP28 secondary structure content significantly. It was not possible to interpret CD spectra of N-lauroylsarcosine-stripped membranes because of the strong chirality of membrane cholesterol, giving negative molar ellipticities at <210 nm wavelength. Taken together, these results suggest that CHIP28 and MIP26 contain 35-55% α–helix. It should be emphasized that CD spectra of membrane proteins can provide only a semiquantitative range of values for secondary structures. A number of concerns exist regarding the selection of basis spectra for membrane proteins, the difference in spectra for transmembrane and external α–helices, the effects of tertiary structure, and the effects of light scattering and absorption.[238]

Because of uncertainties in the interpretation of protein CD spectra as described above, FTIR was used as an independent approach to assess CHIP28 secondary structure. FITR analysis of protein structure is based on the sensitivity of the position and shape of amide absorption resonances to protein secondary structure. The different periodicities and geometries in α–helix vs. β–sheet strongly influence amide hydrogen bonding and thus the absorbance characteristics of the amide resonance. Like CD, a number of assumptions are required in the selection of FTIR basis spectra and in the analysis of data; however, the CD and FTIR methods are complementary because they are subject to very different sources of potential error. In general, FTIR analysis requires the use of partially dehydrated samples to minimize the absorbance of infrared light by water. Although dehydration might influence protein secondary structure, data obtained for a series of pure dried proteins (e.g., myoglobin, lysozyme, bacteriorhodopsin) agree well with secondary structures predicted by crystallographic analysis.

FTIR spectra were obtained on a model 520 Nicolet spectrometer (Nicolet Inc., Madison, WI) with a liquid nitrogen-cooled mercury-cadmium-telluride detector. Measurements were made in the internal reflection mode with an incident beam angle of 45°. Samples of CHIP28 or MIP26 in membranes were air-dried onto the surface of a ZnSe window for spectral measurements from 1700 to 400 cm^{-1}. Figure 6.2 (panel B) shows an FTIR spectrum of stripped-erythrocyte membranes containing only CHIP28 protein. There are two amide resonances—the amide I and amide II peaks. Secondary structure was analyzed by single value decomposition using basis spectra for the amide I peak derived from 17 globular proteins.[196] The amide maxima were at 1650 cm^{-1} and 1551

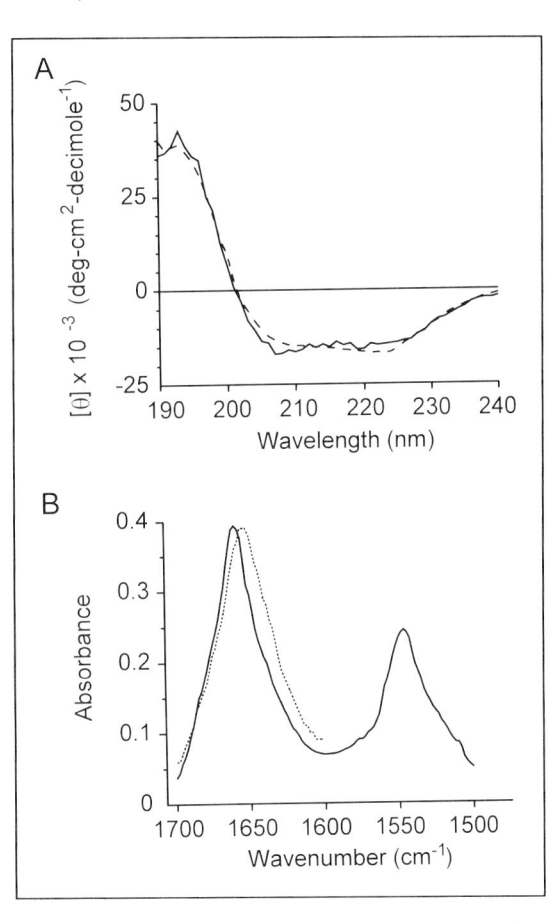

Fig. 6.2. Spectroscopic analysis of CHIP28 secondary structure. A. Circular dichroism spectrum of purified CHIP28 in β–octylglucoside. Spectral decomposition analysis (dashed curve) using the Chang et al set of basis spectra[36] gave ~40% α–helical content. B. Attenuated internal reflection Fourier transform infrared spectroscopy of CHIP28-containing vesicles. Spectral analysis (dotted curve) using the Sarver and Kruger set of basis spectra[196] gave ~45% α–helical content. See text for details.

cm⁻¹; the fitted results (dashed curve) gave 46% α–helix, 37% β–forms and 17% unordered structure. Analysis of FTIR spectra for CHIP28 reconstituted into proteoliposomes gave 46% α–helix, 28% β–forms and 26% unordered structure; analysis of spectra for MIP26 in stripped-lens membranes gave 35% α–helix.[238] These results are in reasonable agreement with the CD data suggesting that CHIP28 and MIP26 have adequate α–helix to form multiple membrane-spanning helical segments.

ANALYSIS OF TRYPTOPHAN ENVIRONMENT BY FLUORESCENCE SPECTROSCOPY

A complementary biophysical approach to examine protein structure is fluorescence spectroscopy—the fluorescence of intrinsic tryptophans[60] or of exogenously-added fluorescent labels at specific sites.[221] A new approach to study intrinsic tryptophan fluorescence has been established recently based on the excitation of fluorescence by beta decay of tritium;[19,199] recent studies of "single photon radioluminescence" of tryptophan residues in CHIP28 support the presence of an aqueous channel that admits only water.[17] CHIP28 contains four tryptophan residues which are predicted by hydropathy to lie at the interfaces between extramembrane and membrane-spanning domains. (See Fig. 5.5, Chapter 5.) Information about the polarity of the environment surrounding tryptophan residues can be obtained from the position of the tryptophan fluorescence emission peak. For example, the emission peak of aqueous tryptophan is at 352 nm, whereas that of azurin, which contains buried tryptophans in a nonpolar environment, is at 308 nm. The tryptophan emission maximum for most proteins is in the range 325-345 nm. Figure 6.3 (panel A) shows fluorescence emission spectra of aqueous tryptophan, CHIP28 and MIP26.[60] Single emission peaks were observed with maxima at 352, 324 and 335 nm, respectively, suggesting a nonpolar environment for tryptophans in CHIP28.

To test the hypothesis that the tryptophan residues in CHIP28 are in a nonpolar environment that is shielded from the aqueous compartment, accessibility of the

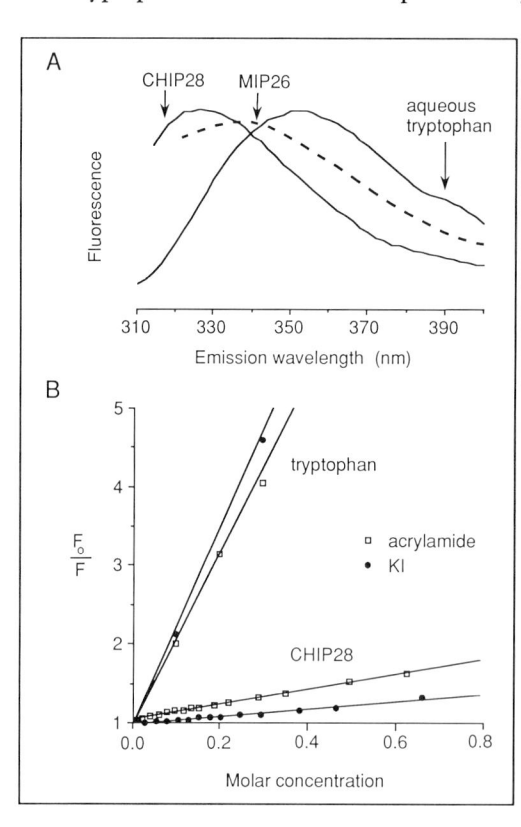

Fig. 6.3. Analysis of intrinsic CHIP28 tryptophan fluorescence. A. Fluorescence emission spectra of CHIP28 and MIP26 in vesicles, and tryptophan in aqueous solution. Note the blue-shift in emission maxima for CHIP28 which indicates a nonpolar tryptophan environment. B. Stern-Volmer plot for quenching of the fluorescence of tryptophans in CHIP28 and in aqueous solutions by acrylamide and potassium iodide.

aqueous-phase quenchers iodide and acrylamide was examined.[60] Iodide and acrylamide quench tryptophan fluorescence by a collisional mechanism in which direct contact must take place. There was very little quenching of CHIP28 tryptophan fluorescence by the aqueous-phase quenchers (Fig. 6.3, panel B); the Stern-Volmer constants, which represent collisional frequencies, were 0.13 M^{-1} (iodide) and 0.71 M^{-1} (acrylamide), much lower than the values of 11 M^{-1} (iodide) and 14 M^{-1} (acrylamide) for quenching of aqueous tryptophan. The shape of the CHIP28 tryptophan emission spectrum did not change with quenchers, suggesting that all for tryptophans are in a nonpolar environment. In contrast, there was significant quenching of MIP26 tryptophan fluorescence by iodide (2.2 M^{-1}) and a blue-shift in emission peak. These results suggest that the nonconserved tryptophan residue near the N-terminus of MIP26 (W2) is in a polar environment.

The spectral and quenching data indicate a nonpolar environment for tryptophans in CHIP28 but do not distinguish whether the tryptophans reside in the membrane bilayer or in a hydrophobic extramembrane domain. To test whether the tryptophans reside in the membrane, effects of membrane-phase quenchers were studied. We used the lipophilic N-anthroyloxy fatty acids (n-AF) in which the anthroyloxy chromophore is positioned at specified depths in the bilayer. It was found that the n-AF probes quenched CHIP28 tryptophan fluorescence by up to 80% with greatest quenching for n=2 and 12. These results support the conclusion that all four tryptophans in CHIP28 are located in a nonpolar, membrane-associated environment; mathematical modeling of the n-AF data suggested that the tryptophans are clustered near the surface and center of the bilayer. Interestingly, the inhibitor $HgCl_2$ was found to quench CHIP28 tryptophan fluorescence by up to 70% with a biphasic concentration dependence; the data suggested at least two sites of action of $HgCl_2$.

CHIP28 Forms Tetramers in Membranes

It is not uncommon for hydrophobic membrane proteins to associate into noncovalent oligomers in membranes. The formation of oligomers may be important for function, e.g., several monomers associate to form an oligomer having a central aqueous pore, or for stability, e.g., hydrophobic monomer-monomer interactions. Several lines of evidence suggested that CHIP28 can associate in oligomers. SDS-PAGE analysis of cross-linked CHIP28[9,214] (using glutaraldehyde[211] or dithiobis(succinimidylpropionate[239]) and sedimentation analysis of detergent-solubilized CHIP28[211] suggested association in dimers, tetramers and/or other higher order oligomers. Size exclusion HPLC of OG-solubilized CHIP28 on a TSK G3000SW column indicated elution primarily as dimers.[238] Mathematical analysis of CHIP28 structure (see above) indicated a number of potential sites for monomer-monomer interactions to yield stable CHIP28 oligomers.

To examine the oligomeric structure of CHIP28 in membranes, we utilized a morphological approach involving freeze-fracture electron microscopy with rotary shadowing.[240] It is well established that certain integral membrane proteins are visible as intramembrane particles (IMPs) in freeze-fracture micrographs, including the vasopressin-sensitive water channel, the proton ATPase and the MIP26 protein.[27,28] In addition, IMPs from a number of tissues have been shown to form regular orthogonal arrays.[7,21,53,294]

The characteristic shape and morphology of the IMPs give information about the state of protein oligomeric assembly; the spatial distribution of IMPs gives information about membrane organization.

Electron microscopy studies were carried out on both artificial and real membranes including proteoliposomes reconstituted with CHIP28, CHO cells stably transfected with CHIP28 cDNA, and kidney tubule plasma membranes.[240] Vesicle/cell pellets or tissues were fixed in 2% gluteraldehyde, cryoprotected in 30% sucrose, and frozen at -150°C. The specimens were fractured at -130°C under 10^{-7} Torr vacuum and shadowed with a ~1.2 nm thick coat of platinum at 45°, followed by 6 nm of carbon at 90°. The replicas were cleaned and mounted on formvar-coated copper grids for observation.

Figure 6.4 shows a series of freeze-fracture electron micrographs of CHIP28 in artificial and native membranes. In N-lauroylsarcosine-stripped

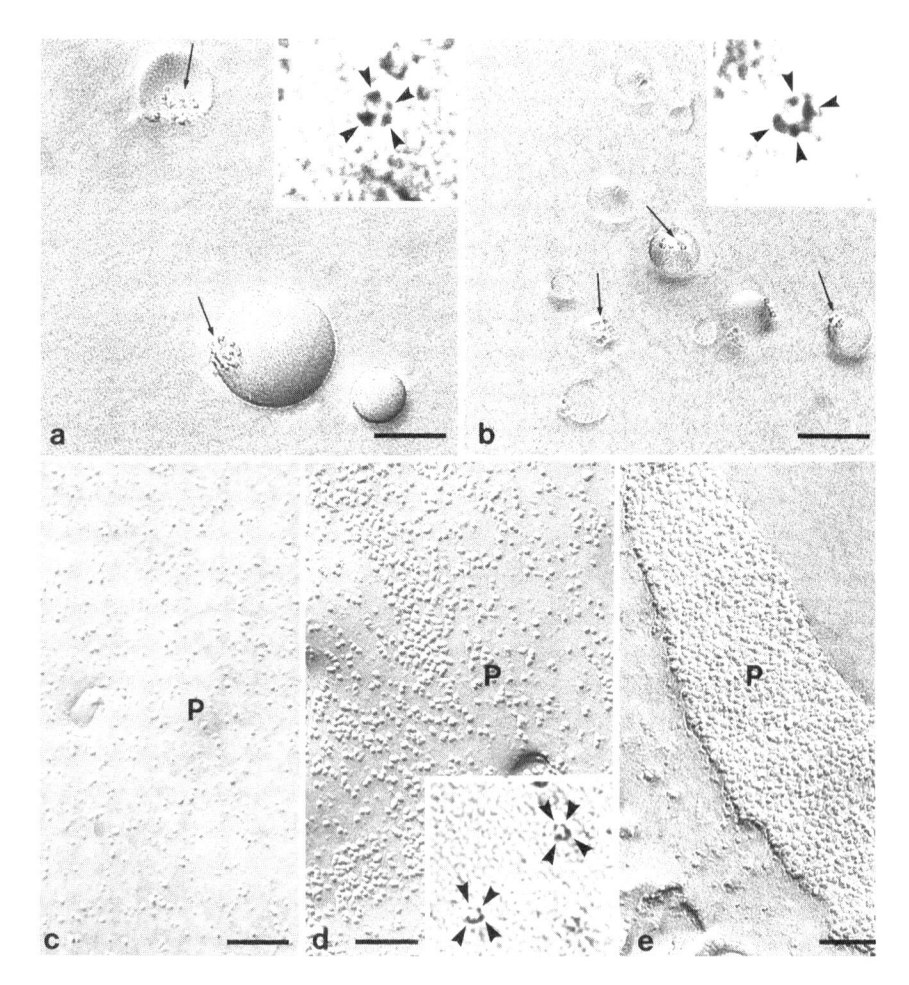

Fig. 6.4. Freeze-fracture electron microscopy of CHIP28. A. N-lauroylsarcosine-stripped erythrocyte vesicles, B. proteoliposomes reconstituted with purified CHIP28, C. mock-transfected CHO cells (P-face), D. CHIP28-expressing CHO cells (P-face), E. Thin descending limb of rat kidney (P-face). Panels A and B (and high-magnification insets) are rotary shadowed and show tetrameric association of CHIP28 monomers. Panels C, D and E are unidirectionally shadowed. Arrows indicate CHIP28 intramembrane particles. Bar = 100 nm. (Figure prepared by J.M. Verbavatz and D. Brown based on the study reported in ref. 240.)

erythrocyte vesicles (panel A) and reconstituted proteoliposomes containing purified functional CHIP28 (panel B), there were a fairly uniform set of IMPs with an average diameter of 8.5 nm. At higher magnification in rotary shadowed specimens, each IMP appeared to be composed of four subunits with an apparent central depression where the platinum shadowing was less dense. Similar IMPs were observed in CHO cells that were stably transfected with CHIP28 cDNA, but not in the nontransfected CHO cells (panels C and D). Panel E shows a freeze-fracture micrograph of the apical membrane of thin descending limb of Henle, showing a very high density of CHIP28 IMPs with the same appearance as those in the reconstituted proteoliposomes. Panels C, D and E show P-face imprints. In addition, distinctive complementary E-face IMP imprints were obtained of CHIP28-containing plasma membranes.[240]

In order to quantify IMP size, the greatest diameters of many hundreds of IMPs were measured for each sample using multiple micrographs. Figure 6.5 shows histograms of IMP diameters. The IMP size distribution in the CHIP28 liposomes was unimodal with a mean diameter of 8.5 nm. In cell membranes, the size distribution was fitted to a bimodal Gaussian function. The density of IMPs was remarkably higher in CHIP28-transfected CHO cells compared to mock-transfected cells; the diameter of the additional population of IMPs with density ~3800 IMPs/μm^2, representing CHIP28, was 8-9 nm. The apical plasma membrane of thin descending limb had a very high density of ~10,000 IMPs/μm^2 with ~8 nm diameter,

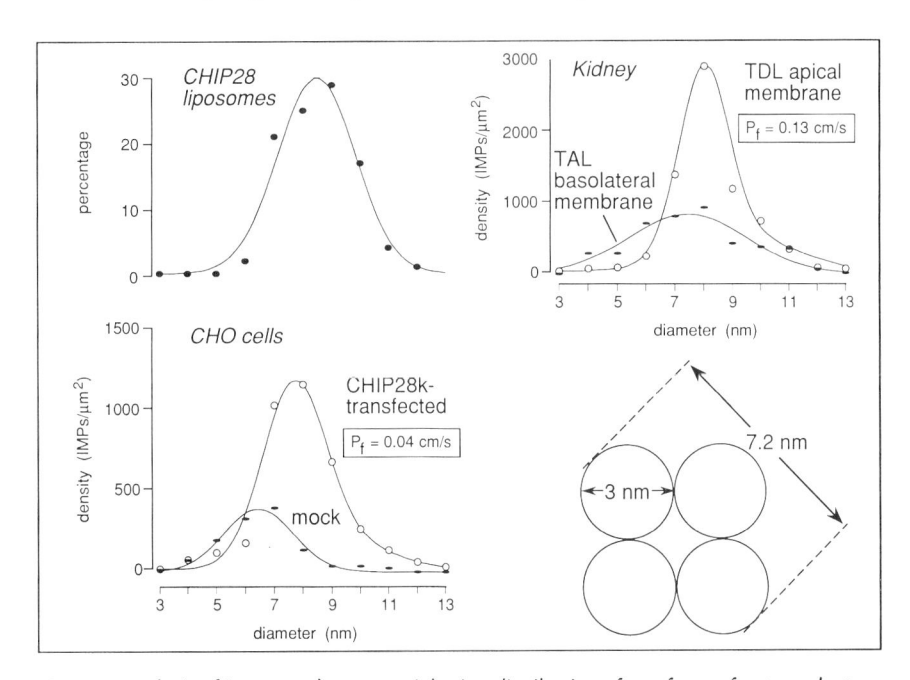

Fig. 6.5. Analysis of intramembrane particle size distributions from freeze-fracture electron microscopy. Samples were proteoliposomes reconstituted with purified CHIP28, CHO cells stably transfected with rat CHIP28 cDNA, control (vector-transfected) CHO cells, and thin descending limb of Henle (TDL) and thick ascending limb of Henle (TAL) from sections of rat kidney. For each sample, diameters were measured for more than 200 IMPs in multiple micrographs; distributions were fitted to Gaussian functions (see text). Lower right: Model for CHIP28 assembly in membrane showing a symmetrical tetrameric assembly of four monomers. (Adapted from ref. 240.)

very different from the distribution of IMPs observed in the water-tight basolateral membrane of the thick ascending limb of Henle. The extraordinary density of 10,000 IMPs/μm^2 in thin descending limb apical membrane suggests that ~22% (w/w) of total membrane protein is CHIP28. For quantitative comparisons, P_f was calculated from the product of the single channel p_f and the water-channel density. For the stably-transfected CHO cells, the product of p_f (~4 x 10^{-14} cm^3/s per CHIP28 monomer at 10°C) and channel density (IMPs/μm^2 x 4, assuming tetrameric association) gives 0.036 cm/s at 10°C, in good agreement with the P_f of 0.027 cm/s measured in CHO cell plasma membrane vesicles. The predicted membrane P_f in thin descending limb of Henle is 0.083 cm/s at 10°C, in agreement with that estimated from published data (transepithelial P_f of 2000 μm/s at 37°C,[115] corrected for apical plasma membrane folding factor of 2.6 and different temperature).

Figure 6.5 (lower right) shows a simple geometric model for a CHIP28 tetramer consisting of a symmetrical arrangement of monomers. If each monomer is a membrane-spanning right cylinder of 5 nm length and protein density 1.3 g/cm^3, then each CHIP28 monomer would have a diameter of 3 nm. The greatest diameter would be 7.2 nm, in good agreement with the measured diameter of ~8.5 nm after correction for the thickness of the platinum shadowing (1-1.5 nm). This calculation supports the conclusion that CHIP28 monomers are assembled in membranes as tetramers. The large central depression observed in the IMPs at high magnification is unlikely to represent the aqueous channel itself because it is very wide (~1.5 nm); estimates of channel diameter based on p_f suggest a narrow channel with diameter ~0.2 nm.[295] In addition, a narrow channel is required to exclude small solutes, monovalent ions and protons as discussed in Chapter 5.

IS THE CHIP28 MONOMER OR TETRAMER THE FUNCTIONAL WATER-TRANSPORTING UNIT?

Determination of the functional water-transporting size of CHIP28 is important to understand the route and mechanism for passage of water. Prior to identification of CHIP28 as the erythrocyte water transporter, radiation inactivation measurements in proximal tubule apical vesicles and erythrocyte membranes gave an ~30 kDa target size for water transport.[235,236] In the radiation inactivation method, frozen membrane samples are irradiated with increasing doses of ionizing radiation. Function is assayed in thawed samples and the "target size" for a selected function is determined from the dependence of functional activity (e.g., water transport) on radiation dose. The target size generally represents the size of the functional protein unit.[128] For a tetrameric association of monomers in which monomer rearrangement does not occur, a target size corresponding to monomer molecular weight would be measured if monomers are functioning independently; a target size corresponding to tetramer molecular weight would be measured if the intact tetramer is required for function or if the monomers are coupled covalently. If the intact tetramer is required for function *and* if rearrangement of monomers occurs after radiation (such that damaged monomers are replaced by functional monomers), then the ln[activity] vs. dose relation would be nonlinear.[261] The finding of a 30 kDa target size for the CHIP28 water channel in native membranes provides good evidence that CHIP28 monomers function independently.

Coexpression studies of mRNA encoding wild-type and mutant CHIP28 proteins in *Xenopus* oocytes provided additional support for independently functioning monomers.[186,296] When *Xenopus* oocytes were injected with mRNA encoding wild-type CHIP28 and a nonfunctional mutant (C189W), oocyte water permeability corresponded to that predicted for the amount of wild-type mRNA injected. It can be concluded that CHIP28 monomers function independently only if the mutant mRNA is expressed on the oocyte plasma membrane and if monomers do not segregate to form functional wild-type tetramers. When *Xenopus* oocytes were injected with mRNA encoding wild-type CHIP28 and a mutant which is expressed normally but does not demonstrate water-transport inhibition by $HgCl_2$ (C189S), it was found that oocyte water permeability was equal to that predicted for the total (wild-type + mutant) mRNA injected and that the percentage water-transport inhibition by $HgCl_2$ was proportional to the fraction of injected mRNA which encoded wild-type CHIP28. In this case, it can be concluded that CHIP28 monomers function independently only if monomers do not segregate to form tetramers consisting of all wild-type or all mutant monomers. Unfortunately, it is not possible in these coexpression studies to rule out the possibility of monomer segregation. By similar reasoning, the demonstration of a linear dependence of P_f on protein-to-lipid ratio in reconstitution studies does not prove independently functioning monomers if, as suggested by freeze-fracture studies, all monomers are assembled as tetramers. Because of the relatively low p_f of single CHIP28 water channels, it is not possible to measure function when only a few (e.g., one to four) monomers are reconstituted into individual liposomes.

A more definitive approach to prove that CHIP28 monomers function independently is to force the formation of oligomers containing both wild-type and mutant CHIP28, based on the approach of Hess and collaborators.[223] If the wild-type-mutant dimeric or tetrameric construct is functional and has activity equal to that of the wild-type protein, then it can be concluded that the monomeric units function independently. A wild type-wild type CHIP28 dimer was constructed by inserting the full CHIP28 coding sequence (contained in a PCR fragment) in-frame just 5' to the stop codon. The 1.6 kb dimer insert encoded a 57 kDa protein, consisting of two CHIP28s in series. Transcribed CHIP28-CHIP28 cRNA was expressed in *Xenopus* oocytes. Water permeability was equal to that obtained when a double quantity of CHIP28 (monomer) cRNA was expressed. Therefore, as in the case of the K channel, the chimeric-dimer was appropriately inserted and targeted. In preliminary studies, dimeric constructs containing wild-type CHIP28 and putative nonfunctional mutants (C189W and NPA box deletions) were functional when expressed in *Xenopus oocytes*, and wild type CHIP28-C189S constructs were functional and showed HgCl2 inhibition intermediate to that obtained by expression of wild-type CHIP28 or C189S monomers (Shi, Skach and Verkman, unpublished results). These results support the conclusion that CHIP28 monomers function independently.

Transmembrane Topology of CHIP28

As described in Chapter 4, the preliminary analysis of CHIP28 hydropathy, taken together with data indicating that the N- and C-termini are cytoplasmically oriented, suggested six membrane-spanning domains.

Analysis of truncated proteins in a cell-free expression system indicated that residue Asn 42, located between hydrophobic regions 1 and 2, is the major site of N-linked glycosylation.[295] Therefore hydrophobic domain 1 spans the membrane with its COOH flanking sequences translocated into the endoplasmic reticulum (ER) at the time of translation. According to the Kyte-Doolittle hydropathy plot, hydrophobic regions 2 and 5 (see Fig. 5.5, Chapter 5) contain fewer than 20 contiguous hydrophobic residues that are believed to be required to span a 4 nm wide lipid bilayer in a standard 3.6-period α–helix.

Polytopic integral membrane proteins in eukaroytic cells that are destined for the cell surface, such as CHIP28, assemble into the ER membrane as the polypeptide chain emerges from the ribosome and rough ER.[144,175,209,276,285] There is strong evidence that the translocation events required to establish transmembrane protein topology are directed by internal sequences in the polypeptide chain which serve to translocate peptide regions across the ER lumen and/or to integrate the chain into the lipid bilayer. Based on these principles of protein translocation, a COOH reporter (the P-domain of prolactin) was used first to map CHIP28 topology. Figure 6.6 illustrates the approach schematically. The P-reporter was chosen because it has been shown to lack intrinsic translocation activity and to respond faithfully to signal and stop transfer sequences in several proteins.[37,208,270] As

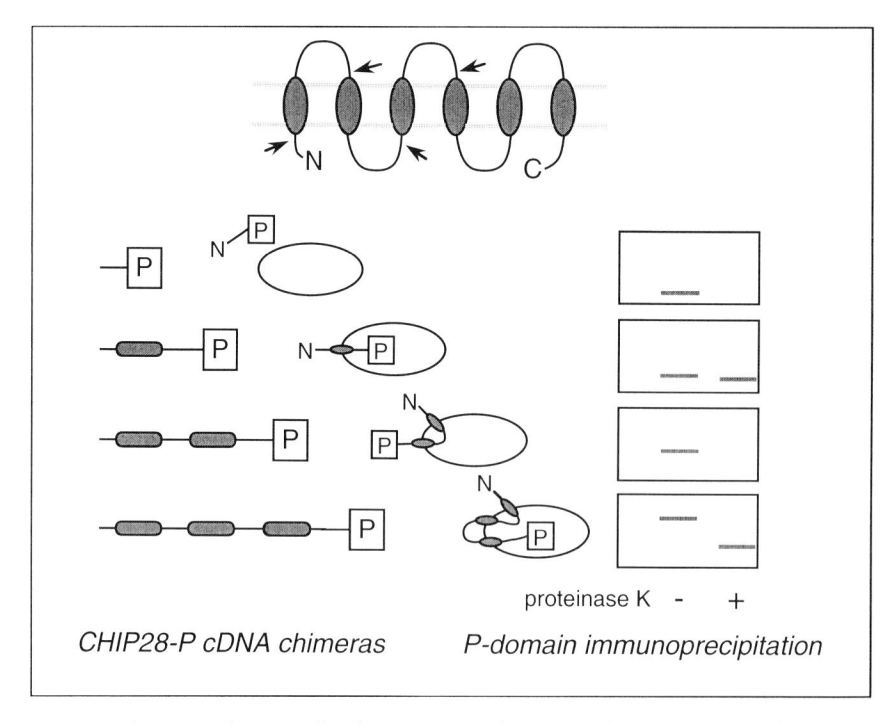

Fig. 6.6. Schematic of strategy for determination of transmembrane topology of polytopic proteins by reporter chimeras. Four possible sites for fusion of CHIP28 to the reporter (P) are shown. If the topology at the top is correct, then the chimeras will insert into the endoplasmic reticulum membrane as shown. The reporter will be proteinase K-sensitive in membranes containing the first and thirds chimeras, and proteinase K-resistant in membranes containing the second and fourth chimeras. See text for details.

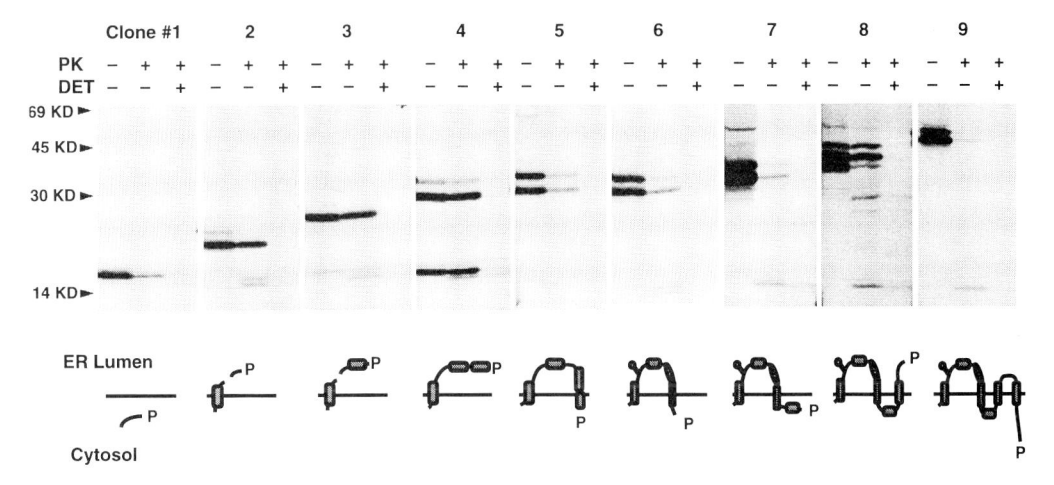

Structure and Water-Transporting Mechanism of CHIP28   71

Fig. 6.7. Determination of CHIP28 transmembrane topology by a reporter chimera. Top: Autoradiograms of translation products generated in Xenopus oocytes injected with CHIP28-P clones and ^{35}S-methionine. Oocyte homogenates were immunoprecipitated with anti-P domain antisera and resolved on SDS-PAGE. Where indicated, proteinase K (PK) or Triton X-100 (DET) were added during oocyte homogenization. The sensitivity/protection of the P-domain epitope to proteinase K digestion is observated by comparison of the first and second lanes of each autoradiogram. Bottom: Predicted topologies for each CHIP28-P chimera based on autoradiograms. See text for details. (Figure prepared by W. Skach based on the study reported in ref. 210.)

described below, the P-reporter approach, taken together with several lines of independent evidence, suggested that CHIP28 spans the bilayer only four times.

To determine CHIP28 topology using the P-reporter, nine chimeras were constructed containing the P-domain fused to a series of putative extramembrane domains in the CHIP28 coding region (Fig. 6.7).[210] Transmembrane topology of the P-domain in ER membranes was determined by accessibility to digestion by proteinase K using P-domain immunoprecipitation and ^{35}S-autoradiography. Resistance to protease indicates translocation into the ER lumen (external orientation), whereas protease sensitivity indicates an extraluminal (cytosolic) location. Figure 6.7 shows an autoradiogram of proteins translated from the nine chimeras. Cell-free experiments (not shown) were performed using rabbit reticulocyte lysate in the absence or presence of pancreatic microsomes. *Xenopus* oocyte experiments were performed by microinjection of cRNAs transcribed from the chimeric cDNAs together with ^{35}S-methionine. Translation of clone #1, containing the first 15 hydrophilic residues of CHIP28 fused to the P-reporter, yielded a peptide of expected size (17 kDa) in oocytes. Proteinase K destroyed P-epitope reactivity, demonstrating that the P-domain was in the cytosol. Proteinase sensitivity was demonstrated for each of the clones by addition of detergent. Cell-free translation of clone #2 (P-domain after hydrophobic region 1) generated a 25 kDa peptide, which was resistant to digestion by proteinase K. Translation of clone #3 (P-domain after hydrophobic region 2) also showed translocation of the P-reporter into the ER lumen. When the P-domain followed hydrophobic domain 3 (clone #4), it was also translocated into the ER lumen; however when the P-domain was placed 14 residues beyond the start of hydrophobic domain 4 (clone #5), it became

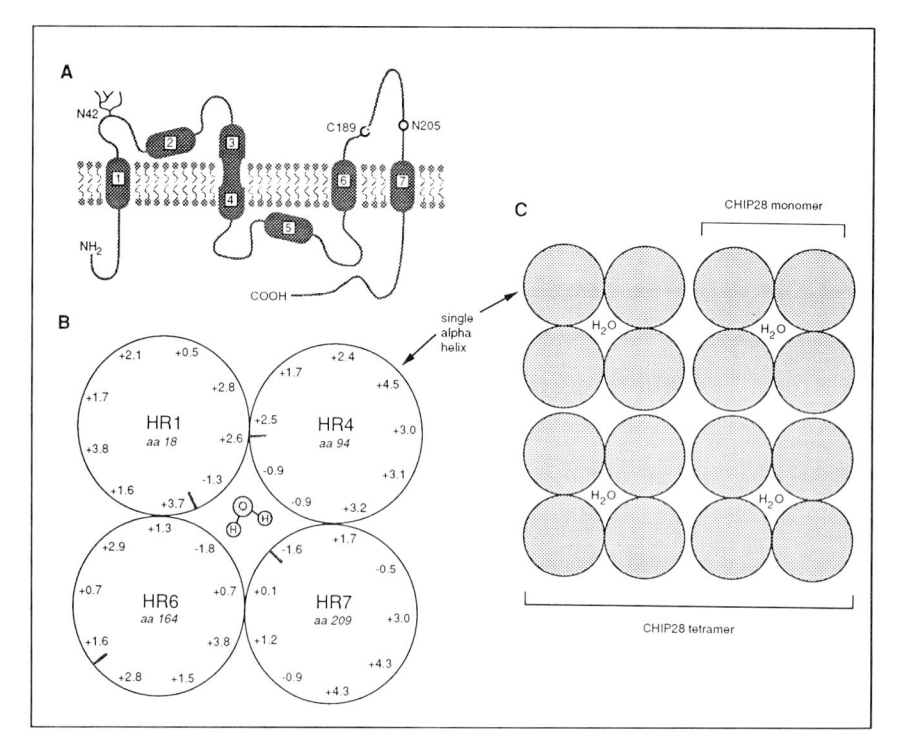

Fig. 6.8. Proposed structure for CHIP28 in membranes. A. Topology showing the seven hydrophobic regions (HR) arranged in four membrane-spanning helices. N-linked glycosylated at N42 and critical cysteine 189 are shown. B. Polarity analysis of the four membrane-spanning helices with Kyte-Doolittle index shown (positive is nonpolar). Note that polar residues line a narrow aqueous channel at the center of the helices. C. Association of functional CHIP28 monomers into tetramers.

proteinase K sensitive. Note the additional bands created by N-linked glycosylation of residue Asn 42. The oocyte data for clones #6-#9 showed resistance to proteinase K only for clone #8. The transmembrane topologies derives from these chimeras are shown beneath the gels. Note that CHIP28 is predicted to span the membrane only four times; two putative transmembrane helices, residues 52-68 and 147-157, reside on the lumen and cytosolic surfaces of the ER membrane, respectively.

Several independent lines of evidence were obtained to support the predicted model shown in Figure 6.8 (panel A).[210] An N-linked glycosylation consensus was engineered at residue 69 by site-directed mutagenesis (H69N). The 6-spanning model predicts that residue 69 is cytoplasmic whereas the model above predicts that it is luminal (external) and thus capable of N-linked glycosylation. It was found that the H69N mutant was able to undergo partial glycosylation. Integration of nascent chains into the ER membrane, as assayed by carbonate extraction,[71] occurred after synthesis of 107 residues and required the presence of two membrane-spanning regions. Internal sequences which initiate and terminate sequential translocation events in CHIP28 biogenesis were identified by engineering CHIP28-derived sequences into defined heterologous chimeras. It was found that: a) a signal sequence within the first 52 residues (containing hydrophobic region 1) initiates nascent chain translocation, b) residues 90-120 encode a stop transfer sequence, and c) residues 126-186 direct translocation of the C-terminus into the ER lumen.

Based on data above supporting the functional independence of CHIP28 monomers, it is predicted that a single aqueous pore should span each monomer. Figure 6.8 (panel B) shows a simple arrangement of four membrane-spanning α–helices (each of ~1 nm diameter) which provide a central pore that can accommodate water molecules. Each helix was rotated to provide a polar pore and a nonpolar external surface. Hydropathy values around the pore (positive indicates nonpolar) are shown.

In summary, the data provided in this chapter suggest that four membrane-spanning α–helices in each functional CHIP28 monomer form a narrow channel for passage of water molecules. Four monomers are associated noncovalently to form tetramers in the membrane (Figure 8, panel C). The proposed model is consistent with functional, topographical, morphological and site-directed mutagenesis data, and provides a simple mechanism by which water traverses the CHIP28 protein. Validation of the model will require high-resolution structural information by electron and/or x-ray diffraction of CHIP28 crystals.[76] Several laboratories are actively engaged in this pursuit. In a collaborative study with Drs. Alok Mitra and Mark Yaeger at Scripps and Drs. Michael Wiener and Alfred van Hoek at U.C.S.F., we have obtained two-dimensional crystals of CHIP28 by reconstitution of detergent-solubilized CHIP28 in artificial lipids. This system should prove useful for examination of CHIP28 structure in native membranes. In a collaborative study with Drs. Michael Wiener, Alfred van Hoek and Robert Stroud at U.C.S.F., we have obtained small crystals of CHIP28 that diffract to low resolution. The conditions to grow crystals that diffract to high resolution are being investigated.

CHAPTER 7

WATER CHANNEL FAMILY RELATIONS

CHIP28 is a member of the MIP26 family of proteins which now includes more than a dozen distinct proteins from mammals, plants, *E. coli* and yeast. The proteins have a tandem repeat motif and share high or absolute amino acid homology in certain regions. Several of the family members have been shown recently to be water channels, and the highly conserved regions have been used to identify new candidate water channels by PCR cloning. This chapter reviews the structure and function of MIP26 family members, as well as homologous cDNAs that have been cloned recently.

THE MIP26 FAMILY OF PROTEINS

A search of the protein data base reveals significant homology (>35% amino identity after sequence alignment) between CHIP28 and several proteins including: MIP26 (major intrinsic protein of lens fiber),[5,55,79,200,206,218] NOD26 (soybean nodulin),[48,161,195] TUR (turgor response protein),[89] TIP (tonoplast intrinsic protein),[155] GlpF (glycerol facilitator protein),[120,166,219] BiB (big brain protein)[188] and ToB (tobacco mosaic protein). In addition, several homologous cDNAs from mammalian cells have been cloned recently (see below).[69,73,104,116,146,148] Figure 7.1 shows an amino acid sequence alignment for ten members of the MIP26 family. Although homologous isoforms for many of these proteins exist, only one form was chosen for sequence alignment and analysis (to avoid biasing the analysis). CHIP28, MIP26, WCH-CD and WCH3 are mammalian proteins, TUR, TIP and NOD26 are plant proteins, BiB is from *Drosophilia* and GlpF from *E. coli*. An homologous protein from yeast was not included because it contained only one NPA consensus sequence (see below).

The sequence alignment in Figure 7.1 shows relatively few internal gaps for proteins derived from very different sources. Based on the tandem repeat motif in which the two halves of each protein are homologous,[172,279] corresponding regions of the first half are shown above those of the second half. Hydrophobic amino acids (*ala, val, phe, pro, met, ile, leu*) are shown in green, charged amino acids (*asp, glu, lys, arg*) in red, relatively polar amino acids (*ser, thr, tyr, his, cys, asn, gln, trp*) in blue and glycine is shown in black. (This color assignment corresponds to the chemistry of the side chains and does not correlate directly with hydropathy indices).[24] Boxes denote absolutely conserved amino acids; note the two conserved "NPA (*asn, pro, ala*) boxes" and the conserved isolated *asp, glu, pro, gln, tyr, his, ser* and *gly* residues. The green segments indicate hydrophobic regions that are potential membrane-spanning domains, and the red segments are potential extramembrane domains. Note that a channel-forming transmembrane he-

Fig. 7.1. Amino acid sequence alignment for ten members of the MIP26 family. Sequences were extracted from the Swissprot data base (except for TUR[89]) and alignment was carried out by the GCG software package. Based on the tandem repeat motif (see text), corresponding regions of the first half are shown above those of the second half. Residue positions are indicated. Hydrophobic amino acids are shown in green, charged amino acids in red, relatively polar amino acids in blue and glycine in black. See text for details. Boxes denote conserved amino acids.

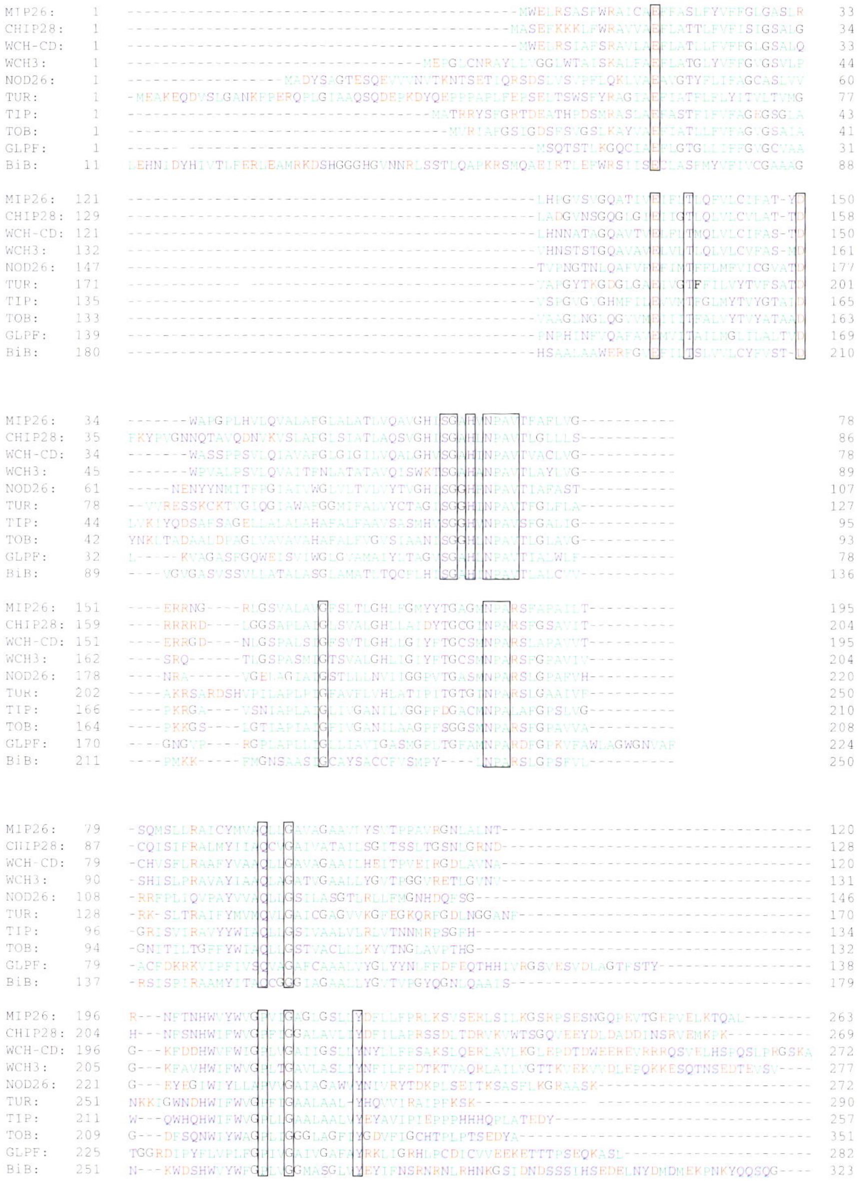

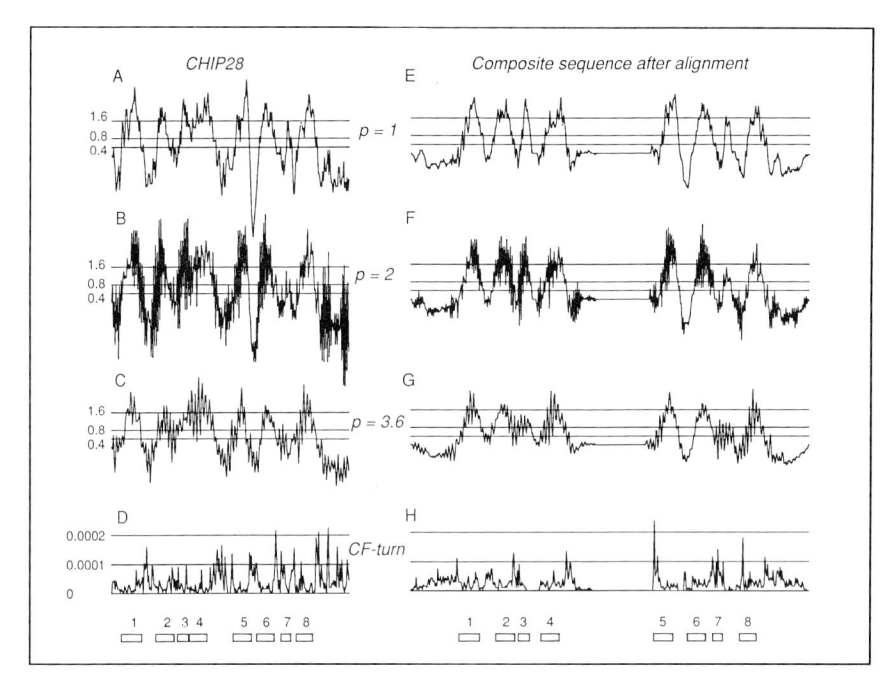

Fig. 7.2. Generalized hydropathy analysis for CHIP28 and a composite sequence obtained by averaging Kyte-Doolittle hydropathy indices for the ten aligned proteins shown in Fig. 1. The analysis was carried out as described in Chapter 6, Fig. 6.1. See text for explanations. (Calculation carried out by A.N. van Hoek).

lix must contain both hydrophobic residues and polar residues which line an aqueous pore. The tandem repeat motif, taken together with data for the genomic structure of MIP26,[82] suggests that MIP26 family members may have evolved from duplication of a single structure representing an ancestral monomer capable of forming homodimers.[172]

Figure 7.2 showed a generalized hydropathy analysis of a composite sequence obtained by averaging Kyte-Doolittle hydropathy indices for each amino acid position after sequence alignment. The analysis was carried out as described for Chapter 6, Figure 6.1 and the curves for CHIP28 are shown for comparison. The gaps in the plots reflect gaps in the sequence alignment as shown in Figure 7.1. Hydropathy analysis of the composite sequence shows up to six membrane-spanning domains with polar N- and C-termini. Regions 4 and 8 (and possibly 1) are potential amphipathic α-helices, and regions 2 and 6 (and possibly 3) are potential amphipathic β-structure. The Chou-Fasman turn prediction identifies most of the interfacial regions between hydrophobic and polar segments.

Further analysis of the composite sequence was carried out by a modification of the hydropathy and variability approach of Donnelly et al.[56] The Fourier transform power spectrum $P(\omega,n)$ was calculated from the equation,

$$P(\omega,n) = [\ \Sigma\ h(j)\ \sin\ (j\omega)\ ]^2 + [\ \Sigma\ h(j)\ \cos\ (j\omega)\ ]^2 \quad \text{(Eqn. 1)}$$

where n is the amino acid position, h(j) is the Kyte-Doolittle hydropathy index of amino acid j, and the sum is carried out from j=n-N to j=n+N

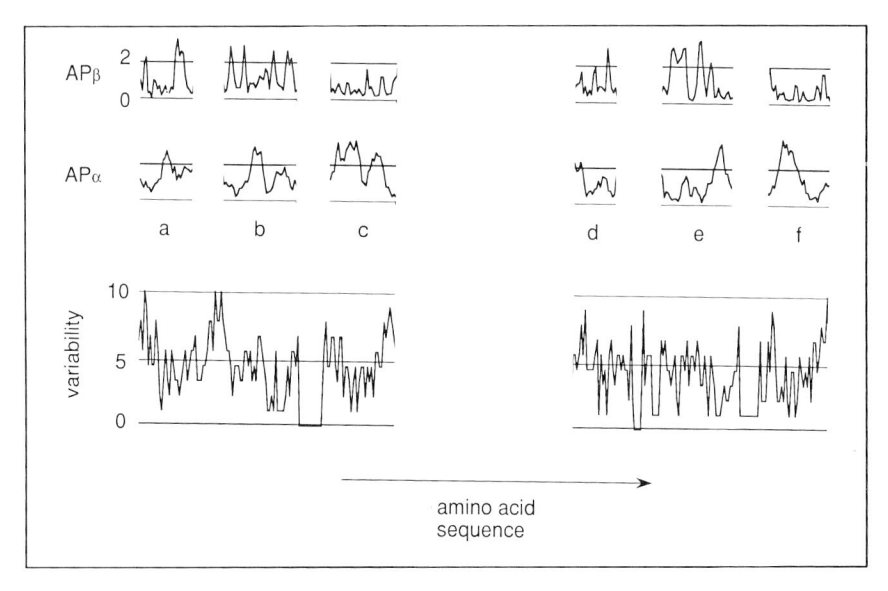

Fig. 7.3. Analysis of the composite aligned sequence of the MIP26 family by amphipathic periodicity (AP) and variability. AP values for α and β structures were calculated as described in the text. AP analysis was carried out for six relative conserved hydrophobic regions. The variability index was defined as the number of different amino acids at each position in the aligned sequence. A lower variability corresponds to higher amino acid conservation. The gap is in the middle of the sequence. See text for interpretations. (Calculation carried out by A.N. van Hoek).

where 2N+1 is the window size. The amphipathic periodicity AP(n) is calculated by integration of P(ω,n) over a restricted angle interval normalized to the integration of P(ω,n) from 0 to 180°. For calculation of α-helix periodicity AP(n)α, the integration is carried out from 90 to 120°. For calculation of β-sheet periodicity AP(n)β, the integration is carried out from 0 to 15° and from 165 to 180°. Figure 7.3 (top) shows the results of the analysis. Amphipathic periodicities are shown for amino acids in each of six conserved regions (a-f) of the composite sequence. Because AP values of >2 are significant, the results suggest the presence of at least four α-helices (regions b, c, e and f). Figure 7.3 (bottom) shows the variability profile, which provides a measure of amino acid conservation for different members of a protein family. A variability value of one indicates abolute conservation and a value of 10 indicates that all of the 10 MIP26 family members have different amino acids at a specified position. The low variability values in the regions b, c, e and f are consistent with a common structural motif; these regions may form the aqueous channel through the water-permeable members of the MIP26 family.

The MIP26 family members have heterogeneous functions. As described in Chapter 5, CHIP28 is a selective water channel. MIP26 is present in very high quantities in plasma membranes in mammalian lens fiber. There is evidence that MIP26 forms relatively nonselective channels that can transport glucose and some small ions,[200] although definitive data are not available. MIP26 is known to interact with calmodulin[79] and undergo glycation.[218] The GlpF protein is a glycerol transporter located in the inner membrane of bacteria.[120] BiB is a neurogenic protein from Drosophilia that functions as an ion channel.[188] BiB is much larger than other proteins in the MIP26 family. TIP is expressed abundantly in protein storage vacuoles in plant seeds[190] and may function as a water channel (see below). The function(s) of NOD26, TUR and ToB have not been established. NOD26 is a plant-encoded protein present in the peribacteroid membrane of soybean root nodules.[161,195] TUR gene transcription increases strongly upon dehydration of pea shoots,[89] suggesting a role in fluid transport. The possibility that some of these proteins can serve as water channels has been investigated as described below.

Water Transport Properties of MIP26 Family Proteins

Because MIP26 family proteins are homologous to CHIP28 and are known to transport small solutes and/or ions,[279] several investigations were carried out to determine whether MIP26 family members were water permeable. MIP26 water transport was studied in reconstituted proteoliposomes and *Xenopus* oocytes.[238,244] In proteoliposomes reconstituted with purified MIP26 from bovine lens, osmotic water permeability was neither sensitive to $HgCl_2$ nor different from that measured in protein-free liposomes. Proteoliposomes reconstituted with CHIP28 and prepared in parallel showed a >50-fold increased P_f. In addition, freshly isolated membrane vesicles from bovine lens, which contained MIP26 as the major membrane protein, had low water permeability. In *Xenopus* oocytes microinjected with 50 ng of cRNA encoding bovine MIP26, P_f was increased by <2-fold above that in water-injected (control) oocytes; oocytes injected in parallel with 5 ng of cRNA encoding CHIP28 had 6- to 10-fold increased water permeability. These results suggest that MIP26 transports little or no water under the conditions of the study. However, because of the negative result, the possibility that MIP26 might become water permeable after certain biochemical modifications cannot be excluded.

Maurel et al[155] reported recently that the TIP protein is water permeable. Expression of γ-TIP in *Xenopus* oocytes increased water permeability by six- to eight-fold. Voltage clamp and isotopic uptake studies indicated that γ-TIP did not transport ions or glycerol. TIP is located in the tonoplast of plant cells and may play a role in vacuole osmoregulation; the expression TIP homologs in peas is induced by water stress. These proteins may thus play a role in the uptake of water in plant roots. Expression of the *E. coli* glycerol facilitator (GlpF)[155] and soybean nodulin protein (NOD, unpublished data, cDNA provided by A. Verma) in *Xenopus* oocytes gave no increase in water permeability, suggesting that these MIP26 family members are not water channels.

PCR Cloning of Homologous Proteins

The highly conserved amino acid sequences in the MIP26 family proteins at and adjacent to the two NPA boxes have been utilized to design degenerate primers for PCR amplification of homologous proteins.[73,148] Figure 7.4 shows a nucleotide and amino acid sequence alignment for several MIP26 family members in the regions of the two NPA boxes. Many of the nucleotides are absolutely conserved and others partially conserved. From this alignment, degenerate sense and antisense primers were designed using nucleotide wobble for partially conserved nucleotides and inosine (I) as a nonselective nucleotide.[148] PCR amplification (30 cycles) was carried out at 55°C annealing temperature using 200 pmol of each primer and size-fractionated cDNA (>1 kb) from kidney or lung as template. Two bands from kidney medulla and lung, and one band from kidney cortex with sizes 380-400 bp were obtained, subcloned and sequenced. From rat and human

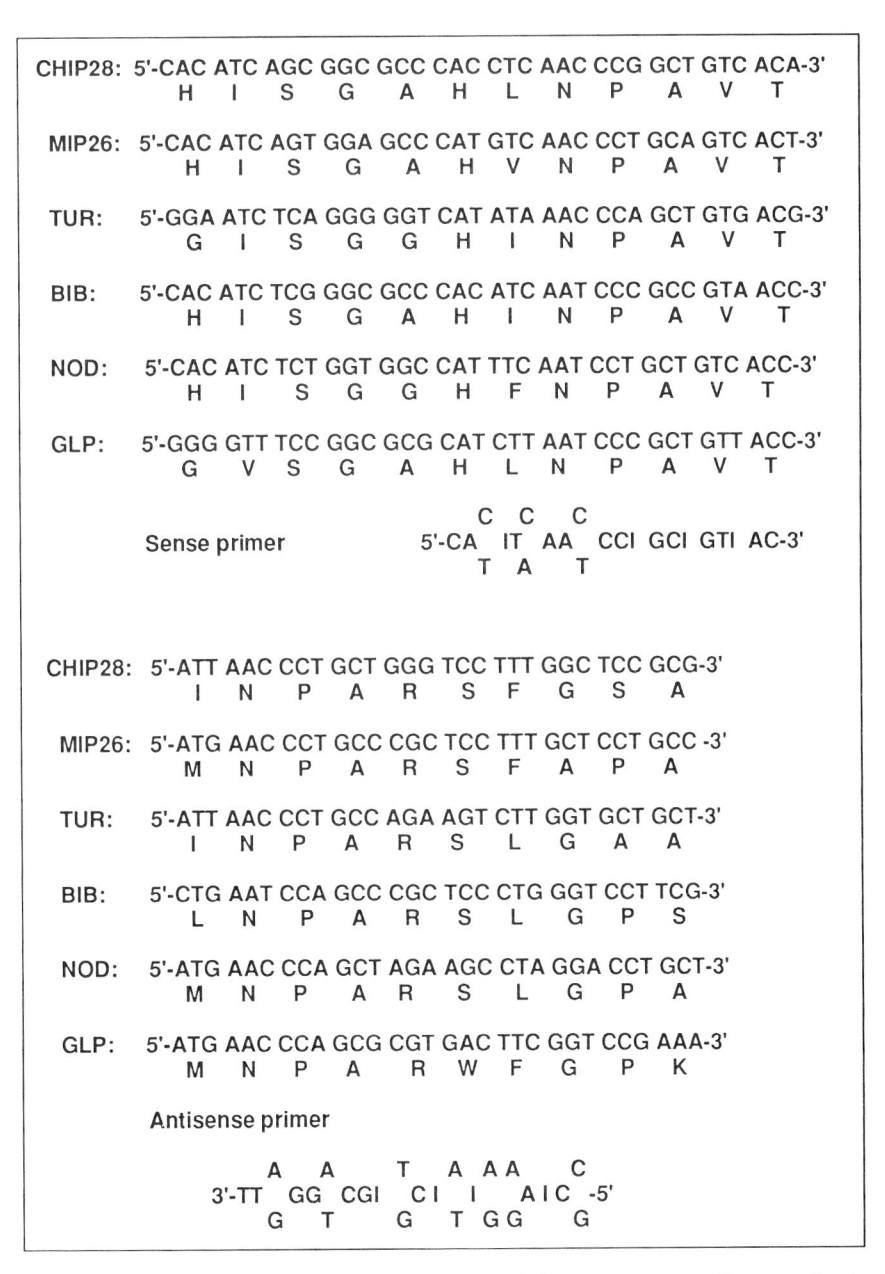

```
CHIP28: 5'-CAC ATC AGC GGC GCC CAC CTC AAC CCG GCT GTC ACA-3'
         H   I   S   G   A   H   L   N   P   A   V   T

MIP26:  5'-CAC ATC AGT GGA GCC CAT GTC AAC CCT GCA GTC ACT-3'
         H   I   S   G   A   H   V   N   P   A   V   T

TUR:    5'-GGA ATC TCA GGG GGT CAT ATA AAC CCA GCT GTG ACG-3'
         G   I   S   G   G   H   I   N   P   A   V   T

BIB:    5'-CAC ATC TCG GGC GCC CAC ATC AAT CCC GCC GTA ACC-3'
         H   I   S   G   A   H   I   N   P   A   V   T

NOD:    5'-CAC ATC TCT GGT GGC CAT TTC AAT CCT GCT GTC ACC-3'
         H   I   S   G   G   H   F   N   P   A   V   T

GLP:    5'-GGG GTT TCC GGC GCG CAT CTT AAT CCC GCT GTT ACC-3'
         G   V   S   G   A   H   L   N   P   A   V   T

                                    C   C   C
Sense primer            5'-CA  IT  AA   CCI GCI GTI AC-3'
                               T   A   T
```

```
CHIP28: 5'-ATT AAC CCT GCT GGG TCC TTT GGC TCC GCG-3'
         I   N   P   A   R   S   F   G   S   A

MIP26:  5'-ATG AAC CCT GCC CGC TCC TTT GCT CCT GCC-3'
         M   N   P   A   R   S   F   A   P   A

TUR:    5'-ATT AAC CCT GCC AGA AGT CTT GGT GCT GCT-3'
         I   N   P   A   R   S   L   G   A   A

BIB:    5'-CTG AAT CCA GCC CGC TCC CTG GGT CCT TCG-3'
         L   N   P   A   R   S   L   G   P   S

NOD:    5'-ATG AAC CCA GCT AGA AGC CTA GGA CCT GCT-3'
         M   N   P   A   R   S   L   G   P   A

GLP:    5'-ATG AAC CCA GCG CGT GAC TTC GGT CCG AAA-3'
         M   N   P   A   R   W   F   G   P   K

Antisense primer
                   A   A   T      A A A      C
         3'-TT  GG  CGI  CI  I     AIC   -5'
                   G   T   G      T GG      G
```

Fig. 7.4. Selection of PCR primers for water channel cloning. Sequence alignment for six members of the MIP26 protein family in the regions of the two conserved NPA amino acid sequences. At positions where nucleotides were not conserved, the primers contained two nucleotides or the nonselective nucleotide inosine (I).

kidney, the CHIP28 sequence and five distinct sequences with 35-55% homology to CHIP28 were obtained. From rat lung, only one homologous sequence was obtained. ^{32}P-labeled DNA probes were used to screen rat and human kidney and rat lung 5'-stretch λgt10 cDNA libraries.

PROPERTIES OF A COLLECTING DUCT WATER CHANNEL HOMOLOGOUS TO CHIP28

Fushimi et al[73] reported recently the cloning of a CHIP28 homolog (WCH-CD) that was expressed exclusively in rat kidney collecting duct. The predicted amino acid sequence of WCH-CD is given in Figure 7.1. WCH-CD is 42% identical to CHIP28 and 59% identical to MIP26. Expression of transcribed cRNA encoding WCH-CD in *Xenopus* oocytes increased water permeability modestly (~3-fold) in a mercurial-sensitive manner. The predicted amino acids sequence showed one potential site for N-linked glycosylation, one potential site for phosphorylation by protein kinase A in the polar C-terminus, and a cysteine residue that corresponds to the mercurial-sensitive cysteine 189 in CHIP28. Northern blot analysis showed expression of mRNA encoding WCH-CD only in kidney and the level of expression was increased in kidneys from dehydrated rats. The mRNA size on Northern blot was 1.5 kb with two weaker bands of greater size. Chromosomal mapping of the human homolog of WCH-CD showed localization to 12q13 which is close to the MIP26 gene.[197] The possibility of defective WCH-CD in non-X-linked diabetes insipidus was raised.

One of the cDNAs isolated by the PCR cloning strategy in our experiments had the same nucleotide sequence as WCH-CD, but was significantly larger at 1.8 kb;[148] Northern blots using probes corresponding to the 5'-untranslated and coding sequences of the 1.8 kb clone showed a single band at ~1.9 kb; expression was increased in kidneys from dehydrated rats (Fig. 7.5). The additional 5'-sequence in the 1.8 kb clone contained an in-frame ATG codon 123 bp proximal to the ATG identified earlier; in vitro translation and oocyte expression studies suggested that the second ATG was the major translation start site. Translation of an engineered WCH-CD cDNA containing a C-terminus c-Myc flag showed a 28-29 kDa protein that was partially glycosylated. Expression of WCH-CD in *Xenopus* oocytes using mRNA contain-

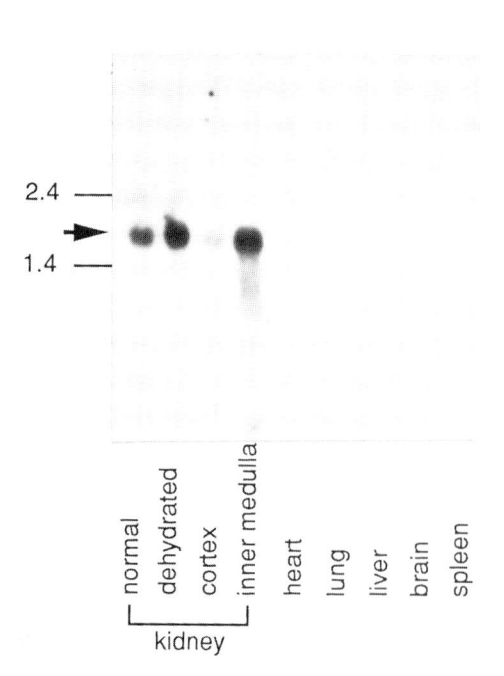

Fig. 7.5. Northern blot of mRNA from kidney and extrarenal tissues probed by a cDNA corresponding to coding sequence of WCH-CD. Where indicated, mRNA was isolated from kidneys from two-day dehydrated rats.

ing a translation enhancer (53 bp 5'-untranslated sequence of *Xenopus* α-globin upstream from coding sequence) gave a large (>10-fold) increase in oocyte water permeability (Fig. 7.6). The incremental water permeability was lower in the absence of the globin sequence (labeled WCH-CD(-)) or when a longer clone containing the upstream ATP codon (labeled WCH-CDL) was expressed. Oocyte water permeability was mercurial sensitive and not increased by cAMP agonists under conditions in which expressed urea[107] transporters and CFTR[105] chloride channels are activated, e.g., 10 min forskolin/IBMX or Sp-cAMPs. Recently, it was shown that extensive incubation with cAMP agonists increased water permeability modestly in oocytes expressing WCH-CD;[132] however, it is unclear whether WCH-CD phosphorylation occurs in native kidney.

As described in Chapter 8, there is evidence based on RT-PCR of microdissected tubules[73] and in situ hybridization[148] that WCH-CD is expressed selectively in kidney collecting duct. Antibody staining suggested WCH-CD protein expression on apical membrane of principal cells[73] and a preliminary immunolocalization study suggested that WCH-CD water channels are present on both apical plasma membrane and subapical vesicles in principal cells.[169]

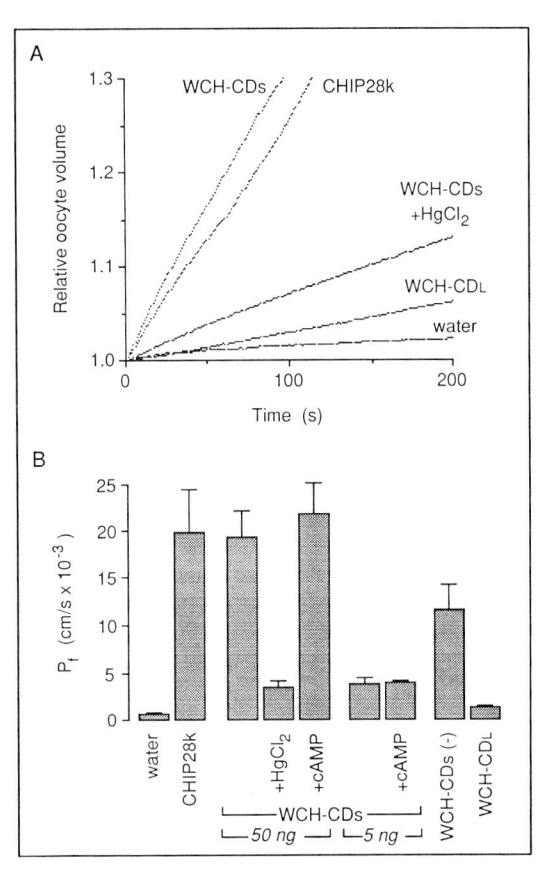

Fig. 7.6. Functional expression of WCH-CD water channels in Xenopus oocytes. A. Time course of swelling in oocytes microinjected with water (control) or 50 ng of transcribed cRNAs encoding rat kidney CHIP28k, the "short" coding sequence of WCH-CD (WCH-CDs) and the "long" coding sequence of WCH-CD (WCH-CDL). Where indicated (+HgCl₂), 0.3 mM HgCl₂ was present. B. Averaged water permeability data (SE) for a series of oocytes. Where indicated (+cAMP), the cAMP agonist Sp-cAMPS (0.1 mM) was added. WCH-CDs(-) indicates that the cRNA did not contain the 5'-untranslated sequence of the Xenopus globin gene.

CLONING OF HOMOLOGOUS PROTEINS

A number of cDNAs with homology to MIP26 have been isolated recently. Frigeri et al[69] identified several proteins from human and rat kidney cDNA libraries. One clone (WCH3, see Figure 7.1 for amino acid sequence) encodes a 276 amino acid protein with 53% homology to CHIP28 and 67% homology

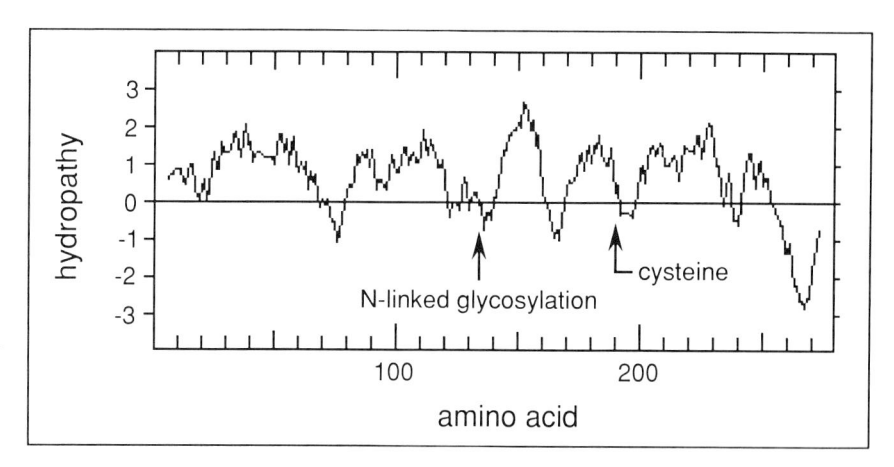

[Figure: Kyte-Doolittle hydropathy plot with hydropathy on y-axis (-3 to 3) and amino acid position on x-axis (up to ~200+), with arrows indicating N-linked glycosylation and cysteine sites]

Fig. 7.7. Kyte-Doolittle hydropathy analysis of protein WCH-3 showing multiple possible membrane-spanning hydrophobic domains, a single N-linked glycosylation site and a cysteine corresponding to cysteine 189 of CHIP28.

to WCH-CD. Hydropathy analysis (Fig. 7.7) shows multiple possible membrane-spanning domains. There is a single consensus site for N-linked glycosylation and a cysteine at position 189. The mRNA of 2.5 kb was expressed selectively in renal papilla > cortex and peptide-derived antibodies indicated protein expression in inner medullary collecting duct cells. Hasegawa et al[104] identified a distinct cDNA that was expressed selectively in rat lung (LAW2, lung alveolar water channel 2) with ~55% homology to CHIP28. Sasaki and coworkers[197] have isolated a cDNA clone from rat kidney encoding a 285 amino acid protein with 34 % homology to CHIP28 and WCH-CD. RT PCR indicated expression in inner medullary collecting ducts, colon, stomach and urinary bladder. Water permeability was increased in oocytes expressing transcribed cRNA. Harris et al[93] have isolated an homologous cDNA from toad urinary bladder with 59% homology to CHIP28. The mRNA is expressed in toad bladder and toad lung, and peptide-derived antibodies detected bands at 32 and 65 kDa. The tissue abundance and physiological significance of these new proteins remains to be determined.

Possible Relationship Between Orthogonal Arrays and Water Channels in Basolateral Membrane of Principal Cells

An unexpected observation made in studies of CHIP28 immunolocalization suggested that a CHIP28/MIP26-like protein is present on the basolateral membrane of principal cells in kidney collecting duct.[224] Polyclonal antibodies against purified human CHIP28 were raised in rabbits. Although the immune serum generally stained only proximal tubule and thin descending limbs of Henle (see Chapter 8), serum from one rabbit also stained strongly the basolateral membrane of principal cells. The serum was named anti-BLIP (BasoLateral Integral Protein). Similar staining in principal cells was seen when the serum was affinity purified against CHIP28 or MIP26. Lens fibers containing MIP26 were stained with BLIP and MIP26 antibodies, but not with CHIP28 antibodies. BLIP antibody also strongly stained basolateral membranes of gastric parietal cells, whereas CHIP28 antibody produced no staining. Immunoblot analysis suggested that the BLIP antibody recognized a 28kDa protein that was distinct from CHIP28. Basolateral membranes of kidney collecting duct principal cells and gastric parietal cells contain abundant orthogonal arrays with appearance similar to that found for MIP26 in lens fibers (see Chapter 8). It was suggested that the BLIP antibody recognized components of orthogonal arrays in basolateral membranes from collecting duct principal cells that may function as water channels. In addition, an anti-OAP (orthogonal arrays of particles) serum against muscle sarcolemmal vesicles stained principal cell basolateral membranes selectively. Protein purification, cloning and expression studies are required to define this potentially interesting relationship between orthogonal arrays and MIP26 family proteins in various tissues.

CHAPTER 8

TISSUE DISTRIBUTION AND PHYSIOLOGY OF WATER CHANNELS

Although it was believed initially that only erythrocytes and kidney tubules contained proteinaceous pathways for facilitated water transport, it is now clear that water channels are distributed widely in selected tissues. The CHIP28 water channel is expressed in many fluid transporting tissues, including kidney proximal tubule and thin descending limb of Henle, choroid plexus, ciliary body, airway epithelium and alveolus, sweat and pancreatic ducts, and others. In contrast, the WCH-CD water channel is only expressed in kidney collecting duct and the new MIP26 family members have yet different tissue distributions. In this chapter, the tissue-specific expression of water channels is reviewed based on blotting, immunolocalization and in situ hybridization experiments. The tissue-specific physiological role of water channels is reviewed and possible nonwater transporting roles of water channels are proposed.

Although low channel-independent (lipid-mediated) water permeability is adequate for regulation of cell volume, high water permeability is required in certain cell plasma membranes. For example, the absorption and secretion of fluids across epithelial cell membranes is an important component of the normal physiology and pathophysiology of many organs. Passive movement of water across cell membranes is driven by osmotic and hydraulic forces which arise from the active transport of ions and osmotically active solutes such as glucose. Rapid movement of water across membranes requires both a suitable driving force and a high membrane water permeability. High membrane water permeability can arise from the presence of functional water channels and/or a large membrane surface area produced by plasma membrane convolutions. The predicted high water permeability in cells containing water channels should thus have important implications for organ-specific mechanisms of secretion/absorption of salt and water.

LOCALIZATION OF CHIP28 AND WCH-CD WATER CHANNELS IN MAMMALIAN KIDNEY

There is a large body of evidence indicating that water permeability is constitutively high in proximal tubule and thin descending limb of Henle, very low in ascending limb of Henle, and regulated by vasopressin in cortical and medullary collecting duct.[246,289] Northern blot analysis of human and rat kidney cortex and papilla showed strong hybridization of a CHIP28 probe at ~2.8 kb[184,295] which in rats was not affected by dehydration.[46,73] Immunoblots of cortical and papilla homogenates with a polyclonal anti-CHIP28 antibody showed a nonglycosylated band at 28 kDa, and a broad glycosylated band whose appearance was tissue- and species-dependent.[103,193] Immunoblots of subcellular fractions from rat kidney probed with anti-CHIP28 antibody were strongly positive for CHIP28 protein in proximal tubule apical vesicles, and weakly positive in proximal tubule basolateral membrane vesicles and purified endocytic vesicles.[193] Northern blot analysis of rat kidney with a WCH-CD probe showed major bands at 1.5 kb and two minor bands of higher size.[73] Similar blots probed with DNAs corresponding to both the 5'-untranslated and coding sequences of WCH-CD showed a single band at ~1.9 kb (see Figure 7.5, Chapter 7);[148] the additional bands in the first study may have arisen from incompletely spliced nuclear RNA. Northern blots showed that the amount of mRNA encoding WCH-CD was upregulated by dehydration.[73,148] mRNA encoding the WCH-CD water channel was not detected in nonrenal tissues.

Immunostaining[103,170,193] and in situ hybridization[20,103,106,295] studies have been performed to examine the tissue localization of CHIP28 water channels in kidney. Immunostaining was carried out by standard methods in which kidneys were fixed in 2% paraformaldehyde, 10 mM Na periodate and 75 mM lysine.[193] Thin sections (1 μm) were cut from frozen tissue blocks on an ultracryomicrotome. The sections were incubated with the primary antibody (immunopurified polyclonal anti-CHIP28 antibody) followed by fluorescent secondary antibody. The washed slides were mounted in 50% glycerol, 0.2 M Tris, pH 8 containing 2% n-propyl gallate for fluorescence imaging.

Immunostaining of rat[170,193] and human[103] kidney cortex with anti-CHIP28 antibody showed strong staining of proximal tubule which was greatest in late S_2 and S_3 segments (Fig. 8.1, top). At high magnification, staining was strongest at the apical brush border membrane, however some staining was detected on basolateral membranes and subapical vesicles. Collecting ducts, glomeruli and other structures in cortex did not stain. Immunogold electron microscopy of the apical plasma membrane of rat proximal tubule and thin descending limb of Henle showed gold particles along the plasma membrane and subapical vesicles (Fig. 8.2, panels c and d) as well as on the basolateral plasma membrane.[193] The gold particles were predominantly located on the cytoplasmic side of the membrane, suggesting that the polyclonal antibodies recognized a cytoplasmic epitope probably near the polar C-terminus. For comparison, immunogold localization of CHIP28 in mock- and CHIP28-transfected CHO cells is shown (Fig. 8.2, panels a and b).

Immunofluorescence micrographs of kidney medulla showed very intense anti-CHIP28 antibody staining of the initial portion of the thin descending limb of Henle, as identified by its continuity with the S_3 segment of proximal tubule, and moderately intense staining of thin limbs down to the papilla (Fig. 8.1, bottom). Ascending limbs of Henle and collecting ducts did not stain; although most vascular structures were negative, it appeared that many vasa recta in the papilla were weakly stained.[193] In thin descending limb of Henle studied at higher magnification, CHIP28 was localized to apical > basolateral plasma membranes.

The conditions for in situ hybridization were optimized based on established procedures.[106] Kidneys from pathogen-free adult male Sprague-Dawley rats (200-250 g) were perfused in situ and fixed in phosphate buffered saline

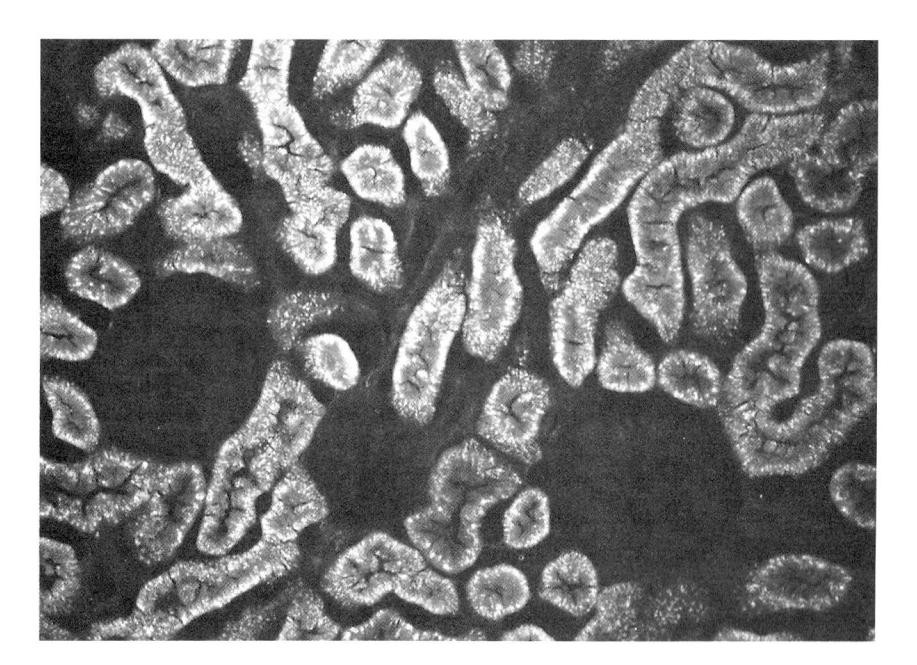

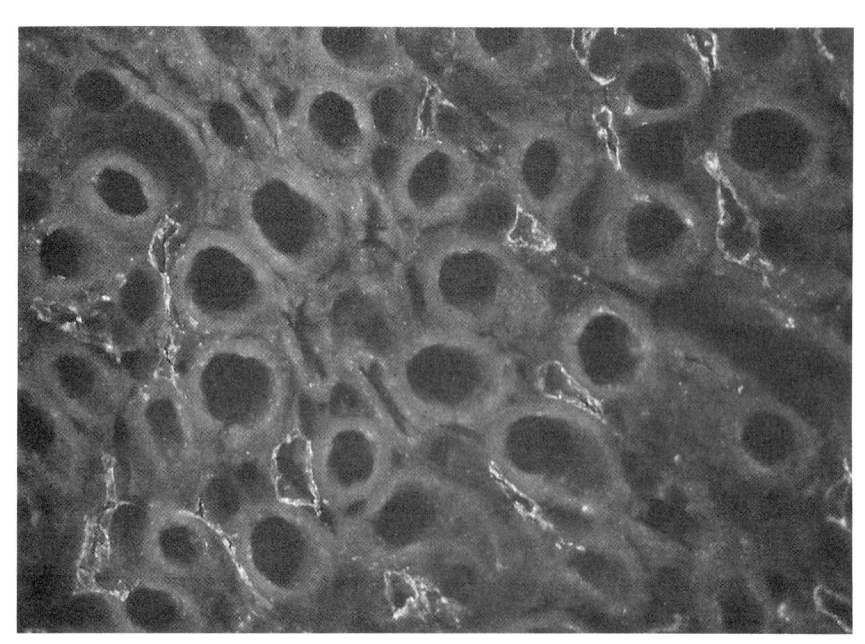

Fig. 8.1. Immunofluorescence localization of CHIP28 water channels in rat kidney with a polyclonal anti-CHIP28 antibody. Top: Section of renal cortex showing localization to proximal tubules. Bottom: Section of renal medulla showing localization to thin descending limbs of Henle.

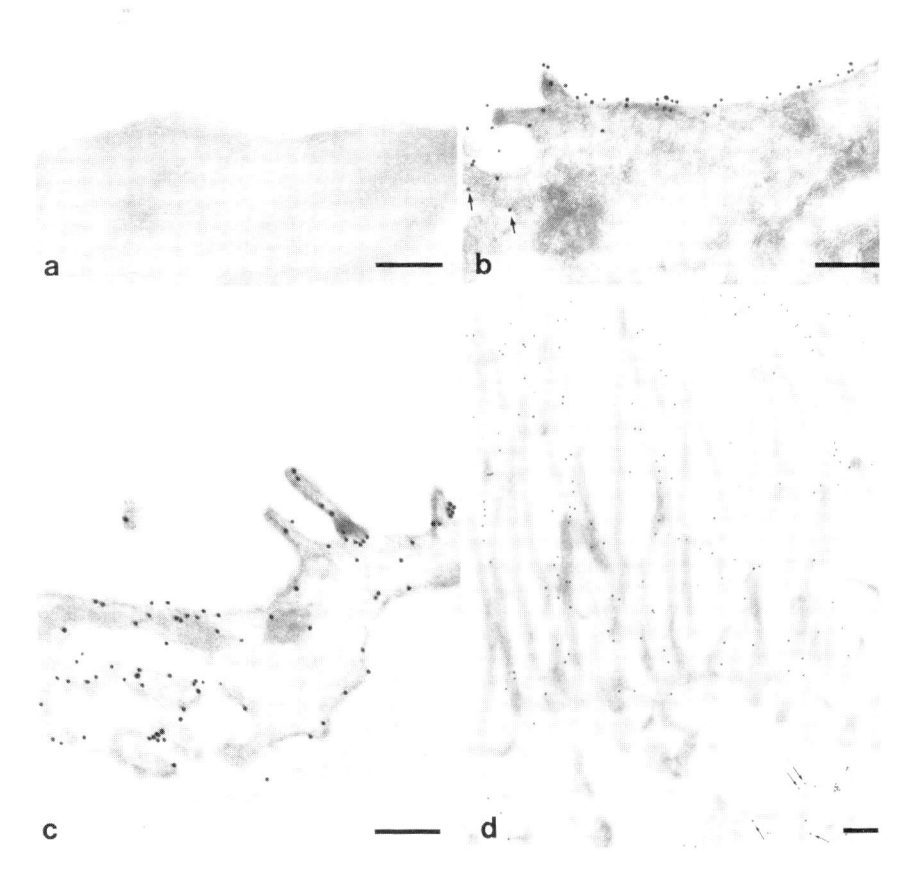

Fig. 8.2. Immunogold detection of CHIP28 in ultrathin sections of transfected CHO cells and rat kidney. a. Mock-transfected CHO cells, b. CHIP28-transfected CHO cells, c. thin descending limb of Henle, d. proximal tubule apical membrane of Brattleboro rat. Arrows indicate immunogold labeling of CHIP28 on plasma membrane and intracellular vesicles. Bar: 250 nm. Photo prepared by J.M. Verbavatz and D. Brown based on studies reported in refs. 147 and 194.

(PBS, pH 7.4) containing 4% paraformaldehyde and 30% sucrose. Kidney slices were embedded in ornithine carbamyl transferase compound, frozen in liquid N_2, and 6-12 μm serial sections were mounted onto coated glass slides. Sections were refixed with 4% paraformaldehyde, rinsed in 0.5x standard sodium citrate (SSC), incubated for 10 min with proteinase K (1 mg/ml in 10 mM Tris, pH 8.0, 0.5 M NaCl), rinsed, treated with acetic anhydride (0.25% in triethanolamine buffer) for 10 min, rinsed, and hybridized overnight at 58°C. Hybridization was performed in 50% deionized formamide, 0.3 M NaCl, 10% dextran sulfate, 20 mM Tris HCl (pH 8.0), 5 mM EDTA, 1.5x Denhardt's solution, 5 mg/ml tRNA and 100 mM dithiothreitol. Approximately 6×10^6 cpm of cRNA probe was applied to individual slides. Slides were rinsed in 2x SSC containing 1 mM EDTA and 10 mM β-mercaptoethanol, treated with 20 μg/ml ribonuclease A, 10 mM Tris (pH 8.0) and 0.5 M NaCl, rinsed, then washed in 0.1x SSC containing 10 mM β-mercaptoethanol and 1 mM EDTA for 1 hr at 57°C. Slides were rinsed in 0.5x SSC and dehydrated in a series of alcohol solutions containing 0.3 M ammonium acetate. Slides were then dipped in Kodak NTB-2 autoradiographic emulsion at 42°C, exposed for six days at 4°C, and developed in Kodak D-19 developer at 16°C for 3 min.

In situ hybridization with an antisense 155-base cRNA probe (corresponding to bp 7-162 of rat CHIP28) showed strong hybridization over proximal tubule epithelial cells (Fig. 8.3, panels A and B) with greatest signal in deep cortex corresponding to distal S_2 and S_3 segments (Fig. 8.4, panel D).[295] The stronger hybridization in proximal tubule cells of the deep vs. superficial cortex is consistent with the higher water permeabilities of proximal straight vs. convoluted tubule,[15] as well as with vesicles studies that indicate the water permeability in the deep cortex is higher than in superficial cortex.[228] There was no hybridization to glomeruli and collecting ducts. In renal medulla, there was hybridization to a subset of tubules corresponding to thin limbs of Henle (Fig. 8.3, panels D and E; Fig. 8.4, panels E and F). There was no specific hybridization of a sense (control) cRNA probe (Fig. 8.3,

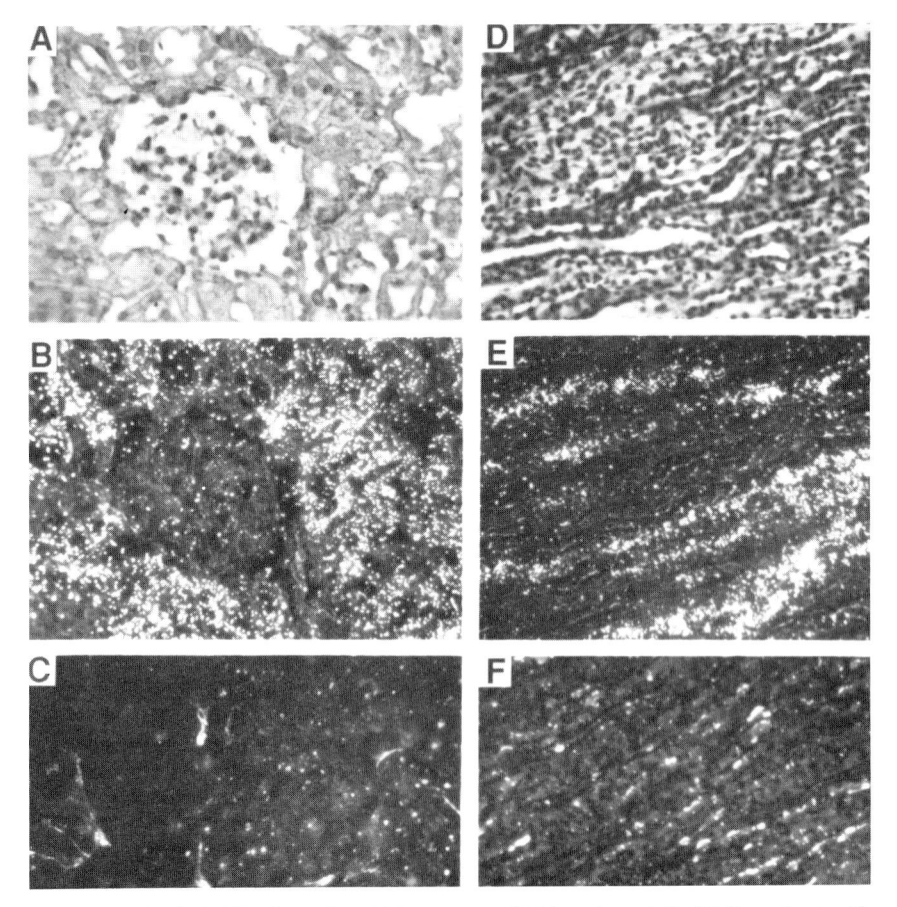

Fig. 8.3. In situ hybridization of rat kidney cortex (A-C) and medulla (D-F) probed with antisense (A, B, D, E) or (control) sense (C, F) CHIP28 cRNA probes. In cortex, areas of positive hybridization in brightfield (black granules, panel A) and darkfield (white dots, B) micrographs correspond to proximal tubule epithelial cells. In papilla, positive hybridization was in thin limbs of Henle. (From ref. 295.)

Tissue Distribution and Physiology of Water Channels

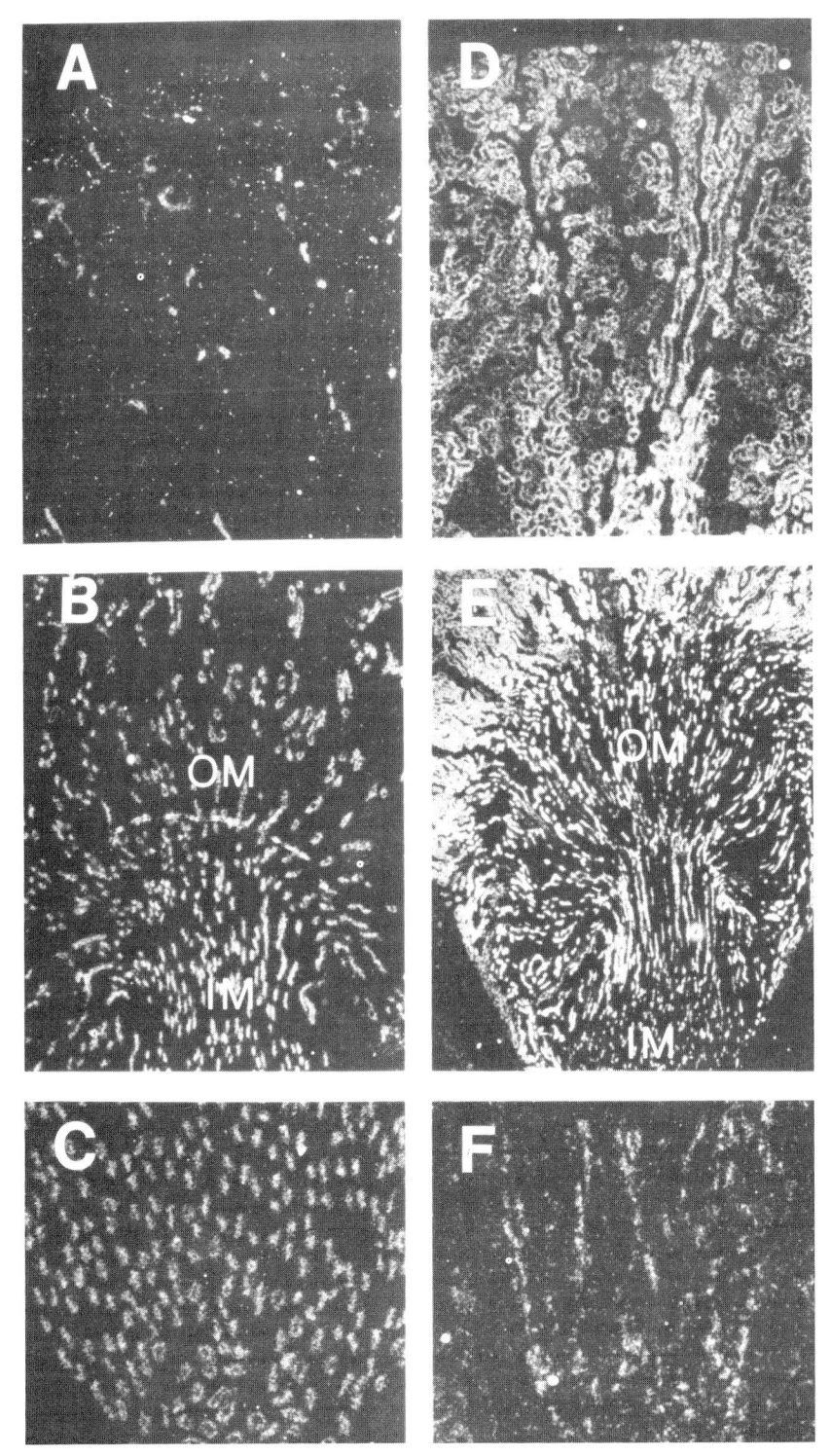

Fig. 8.4. Low magnification in situ hybridization of rat kidney probed with antisense WCH-CD (panels A-C) and CHIP28 (panels D-F) probes. A complete section through the kidney is shown which includes cortex (panels A and D), outer medulla (labeled OM) and inner medulla (labeled IM). (From ref. 148.)

panels C and F). These data are consistent with the results obtained by immunostaining.

The WCH-CD water channel had a very different distribution in the kidney as shown by in situ hybridization (Fig. 8.4, panels A-C).[148] There was relatively weak hybridization of an antisense WCH-CD cRNA probe in cortex to a subpopulation of tubules corresponding to cortical collecting ducts. In medulla, there was hybridization to collecting ducts but not to thin limbs of Henle. Localization of mRNA encoding WCH-CD to collecting ducts was also demonstrated by PCR of microdissected kidney tubules.[73] These results are consistent with antibody staining data showing that a polyclonal antibody raised against a C-terminus peptide of WCH-CD stained the apical plasma membrane of rat collecting duct principal cells.[73]

These results indicate that CHIP28 water channels are expressed in constitutively water permeable nephron segments, whereas WCH-CD water channels are expressed in the vasopressin-sensitive collecting duct. Little or no antibody staining was observed in basolateral membranes of collecting duct and proximal tubule, suggesting the existence of additional water channel(s).

CHIP28 Localization in Nonrenal Tissues

A summary of the tissue distribution of CHIP28 water channels is shown schematically in Figure 8.5. CHIP28 water channels were found selectively in epithelial cells believed to be involved in fluid absorption and/or secretion, and in selected endothelia that are probably involved in transcapillary water transport. In the discussion below, the principal findings are presented and the physiology and evidence for fluid transport in CHIP28-containing extrarenal tissues is reviewed.

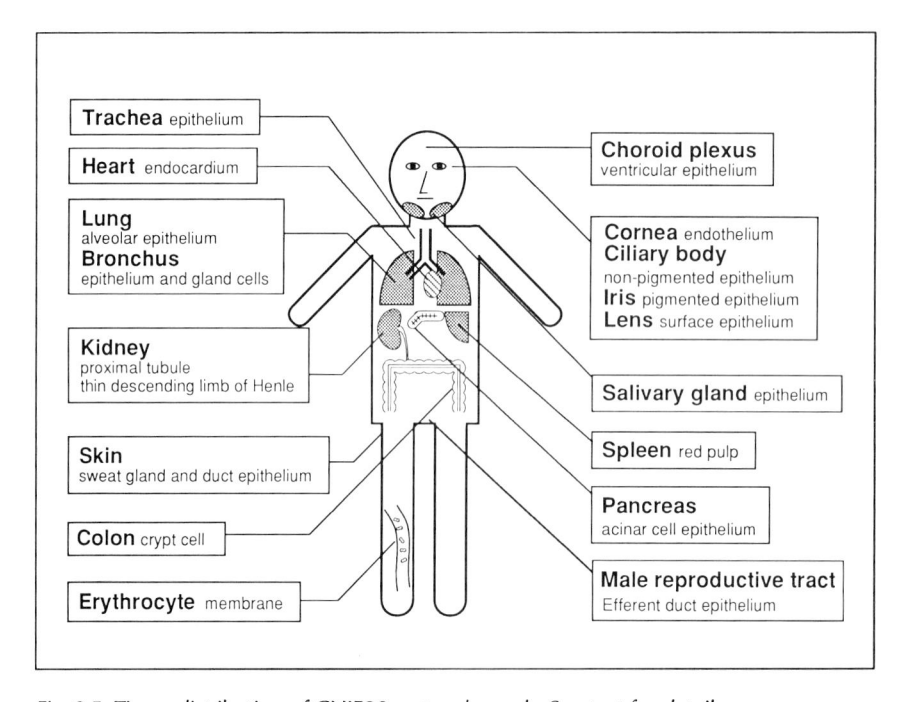

Fig. 8.5. Tissue distribution of CHIP28 water channels. See text for details.

The strongest expression of CHIP28 mRNA and protein was found in choroid plexus.[20,103] Figure 8.6 (top) shows in situ hybridization and Figure 8.7 (top) shows immunostaining of rat choroid plexus. There was intense expression of CHIP28 mRNA and immunoreactive protein in epithelial cells in choroid plexus without expression in other regions of brain. Antibody staining indicated that CHIP28 was localized primarily on the epithelial cell ventricular surface membrane. In this and subsequent figures, hybridization with a (control) sense cRNA probe was negative, and staining with control serum (both preimmune serum and anti-CHIP28 antibody depleted serum obtained by immunoadsorption) was negative.

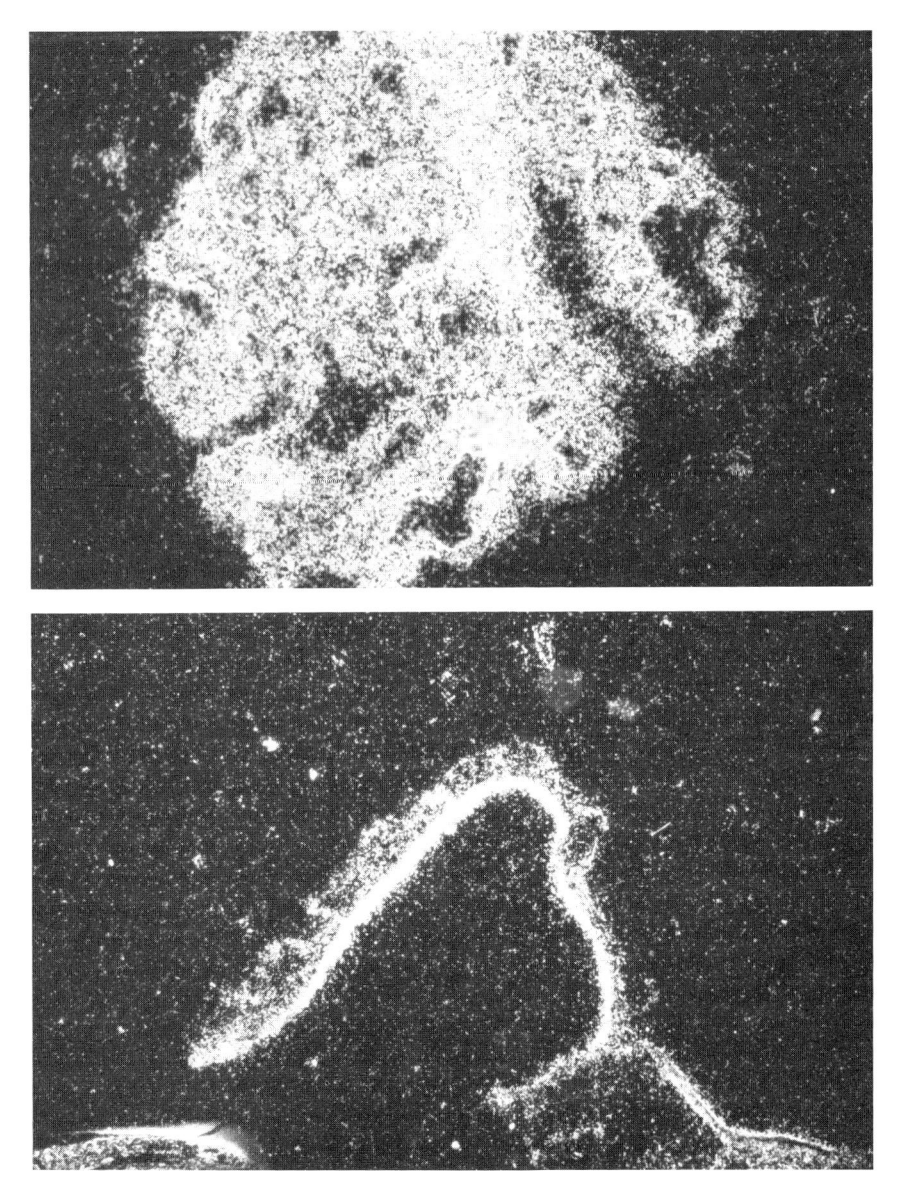

Fig. 8.6. In situ hybridization of rat choroid plexus (top) and iris (bottom) with antisense CHIP28 probe.

Choroid plexus is a secretory epithelium lining brain ventricles and is responsible for the daily secretion of ~500 ml of cerebrospinal fluid (CSF). Fluid secretion is driven by ouabain-sensitive sodium transport. Recent studies showed a correlation between CSF secretion and KCl movement and cAMP-dependent chloride transport.[51] Ion transport is thought to establish a transmembrane solute gradient to drive transcellular osmotic water movement.[291] The CHIP28 water channels on the ventricular (apical) surface of choroid plexus probably facilitate the osmotically-driven movement of water from blood to the cerebrospinal fluid space.

CHIP28 was expressed throughout the small and large airways and the lung.[103,106] Immunolocalization and in situ hybridization studies indicate expression in alveolar epithelial cells, in epithelial and adventitial layers of trachea, and in human submucosal gland serous cells.

The alveolar epithelium consists of two types of cells: squamous type I cells and cuboidal type II cells. Type I cells cover >97% of the alveolar surface but only comprise one-third of the total number of alveolar epithelial cells. Type II cells are located at the "corners" of alveoli and cover only ~3% of the alveolar surface. The localization data suggest that both cell types express CHIP28, with possibly more expression in type II cells. There is good evidence that active solute and fluid transport occurs in the type II cells. Reabsorption of isotonic fluid from the alveolar space is inhibited by ouabain and amiloride.[8] Na/K pumps were immunolocalized in basolateral membranes of type II but not type I alveolar epithelial cells. In addition, there is direct experimental evidence (see below) that CHIP28 water channels probably provide the route for osmotically-driven water movement in intact lung.[67]

The tracheobronchial epithelium is involved in fluid transport for maintenance of mucosal hydration during respiration and for formation of the protective mucous layer. The tracheal epithelium is composed of ciliated and goblet cells of surface epithelium, and serous and mucous cells of submucosal glands. CHIP28 protein was located in apical membrane of the surface epithelial cells and in basolateral membrane of submucosal gland epithelial cells. In trachea, transepithelial chloride efflux from blood to lumen is mediated by electroneutral Na-K-Cl cotransport at the basolateral membrane and cAMP-dependent chloride conductance at the apical membrane.[90,274] Studies of cultured submucosal glands demonstrate active chloride transport, and CFTR is abundant in serosal but not mucosal gland cells.

There was strong expression of CHIP28 mRNA and protein in selected cell types in mammalian eye. There was strong CHIP28 expression in corneal endothelium, and epithelium in iris (Fig. 8.6, bottom; Fig. 8.7, bottom), ciliary body and lens (Fig. 8.7, middle).

The cornea consists of several distinct cell layers with an endothelial cell layer in contact with the anterior chamber. The corneal endothelium transports fluid from the stromal compartment to the space containing aqueous humor. Transport is inhibited by stilbenes, ouabain, amiloride, and the water channel inhibitor pCMBS.[168] CHIP28 water channels probably provide the route for solute-driven water transport and may be important in the maintenance of lens clarity.

The ciliary body produces and secretes aqueous humor, a clear hypotonic fluid with composition similar to that of CSF. The production of

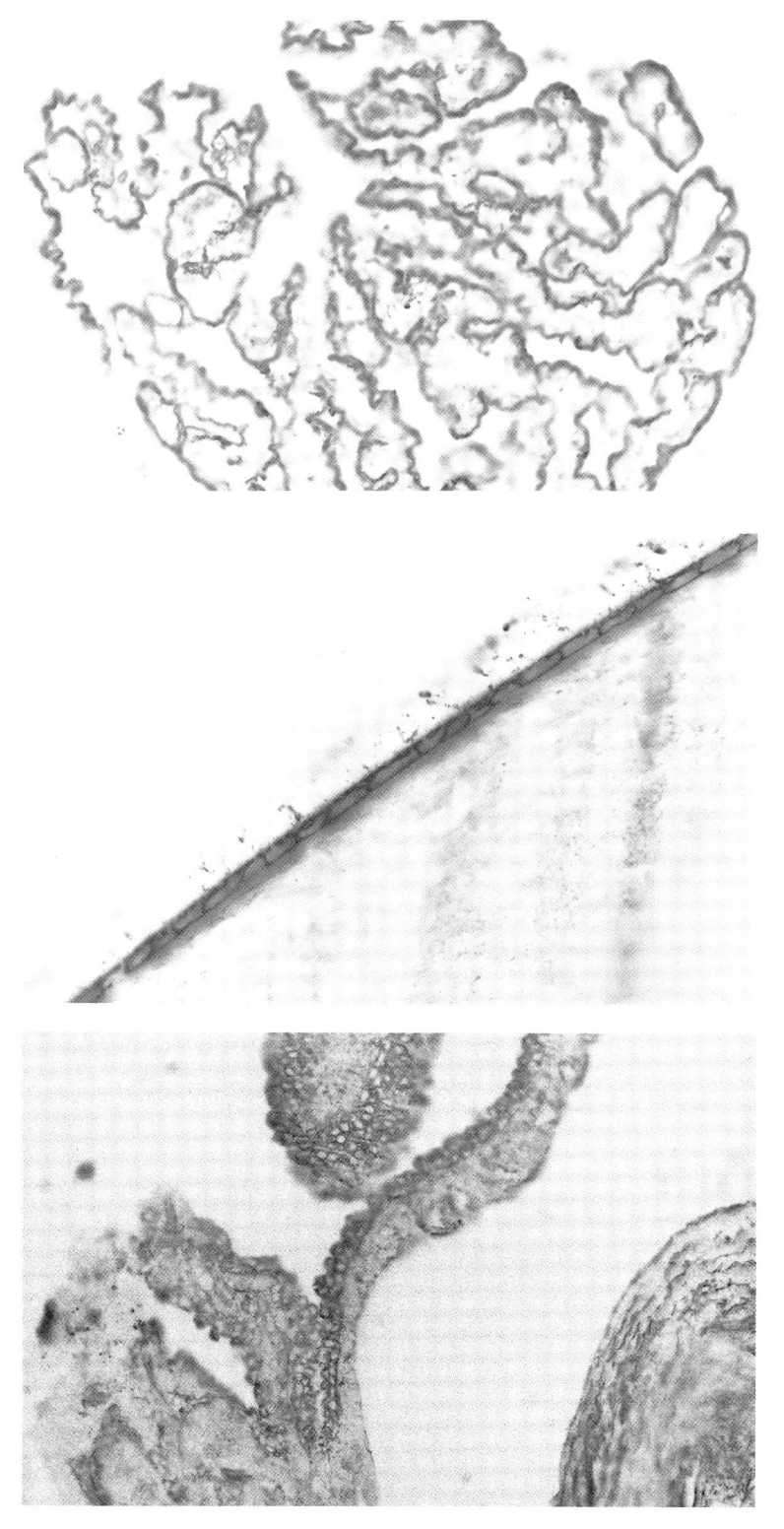

Tissue Distribution and Physiology of Water Channels 93

Fig. 8.7. Immunofluorescence of rat choroid plexus (top), lens (middle) and iris (bottom) with anti-CHIP28 antibody showing strong staining of epithelial cells.

aqueous humor is an active process mediated by pigmented and nonpigmented epithelial layers lining the ciliary process and facing the posterior chamber. Branching epithelial folds in the ciliary body (ciliary processes) are responsible for the continuous production of aqueous humor. CHIP28 water channels were found primarily at the inner surface of ciliary body epithelium facing the posterior chamber, suggesting localization to nonpigmented > pigmented epithelial cells. The nonpigmented cells contain Na/K pumps and Cl/HCO3 exchangers. Facilitated transepithelial ion transport generates the driving force for passive osmotic water movement.

The iris primarily functions as a light diaphragm, and consists of loose vascular connective tissue. Its posterior surface is lined by pigmented and nonpigmented epithelial cell layers. In situ hybridization showed a strong signal over the posterior surface (Fig. 8.6, bottom). Antibody staining was also strongest in the posterior layer. Little is known about ion or fluid transport in the iris epithelium. The presence of CHIP28 water channels suggests that the iris has an important role in fluid transport. CHIP28 was also found in a single layer of surface epithelial cells in lens (Fig. 8.7, middle). The function of these cells is not known, but as for corneal endothelial cells, the lens epithelial cells may be involved in fluid-transport for maintanence of lens transparency.

A series of localization studies were carried out in tissues from the gastrointestinal tract.[103,106] No CHIP28 mRNA or protein was detected in liver, stomach and small intestine. There was low expression in esophageal and salivary glands. The only tissues with clear-cut CHIP28 expression were epithelial cells in colonic crypts and pancreatic acini. Figure 8.8 shows hybridization of the antisense CHIP28 cRNA probe selectively to crypt epithelial cells in rat colon.

Fluid transport in mammalian colon is required for dehydration of feces. Colonic epithelium undergoes continuous cell migration from crypt to villus. Crypt cells are believed to be mainly secretory and villus cells absorptive.[130,275] Recent studies suggest that crypt cells provide an important route for fluid absorption in normal physiology, but may become secretory when stimulated by secretagogues.[156,167] The specific localization of CHIP28 water channels is consistent with the physiological data. The exocrine pancreas is composed of the pancreatic acinus for secretion of alkaline enzyme-rich fluid, and pancreatic duct for excretion of pancreatic fluid into the duodenum. The CHIP28 water channel in pancreatic acinar cells may be important for the secretion of near isosmotic fluid.

Immunostaining of human skin showed CHIP28 localization to both the secretory portion and excretory duct.[103] There was strong staining of basolateral membrane of gland cells and apical staining of duct cells. Sweat glands are widely distributed in human skin and function as a thermoregulatory mechanism by secreting watery fluid onto the skin surface. Except for skin in the axilla and genital regions, ordinary skin sweat gland is a merocrine (eccrine) gland that is regulated by cholinergic sympathetic nerves. These sweat glands are composed of a coiled secretory portion and tubular excretory portion (duct). The secretory portion consists of a single cell layer that transports sodium ions into the gland lumen. This secretory process is accompanied by passive diffusion of water which is probably mediated by CHIP28. The excretory sweat duct is composed of double layers of epithe-

Tissue Distribution and Physiology of Water Channels 95

lial cells. The duct epithelium is thought to reabsorb sodium ions to make sweat hypotonic with respect to plasma.

In an immunolocalization study of male reproductive tract,[31] CHIP28 was expressed in brush border and basolateral plasma membranes of nonciliated cells in efferent duct. There was also staining of epithelial cells in the ampulla of vas deferens, seminal vesicles and prostate, but not the cells in seminiferous tubules, epididymis and the proximal vas deferens. It

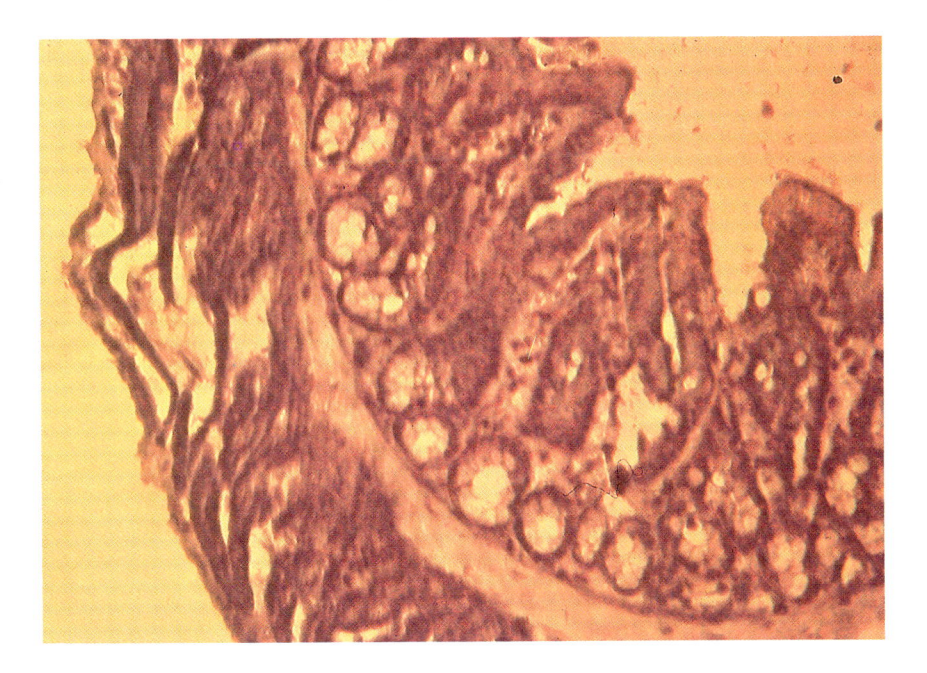

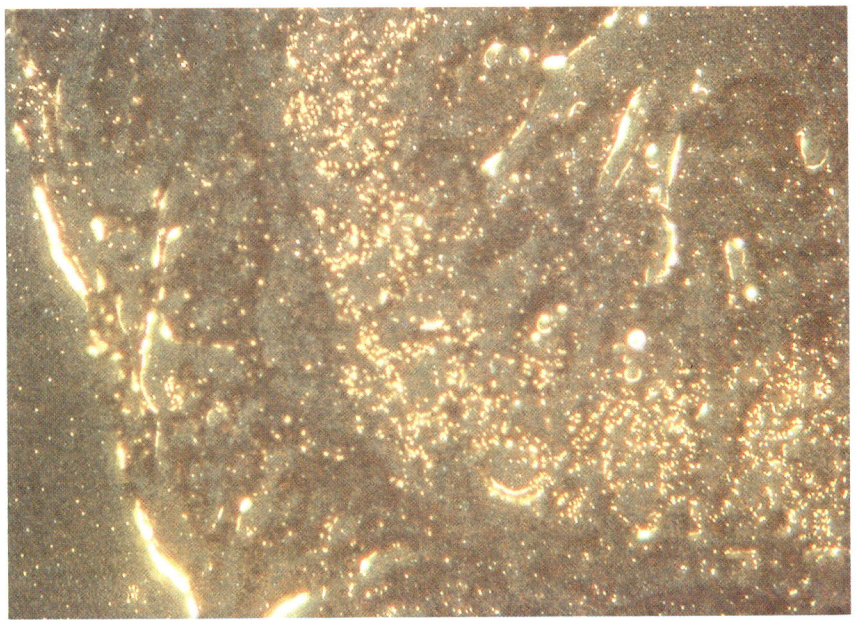

Fig. 8.8. In situ hybridization of rat colon with antisense CHIP28 probe showing stained (top) and darkfield (bottom) sections. Hybridization was observed on crypt epithelial cells.

was proposed that CHIP28 is a principal mediator of transmembrane water transport in absorptive epithelial cell in the efferent duct.

Hybridization studies in spleen showed localization of CHIP28 mRNA in red splenic pulp.[106] The red splenic pulp contains erythroid cells in various stages of development, whereas the white pulp contains lymphoid and myeloid elements. The localization of mRNA encoding CHIP28k in erythroid elements is consistent with the abundant quantities (1-2 x 10⁵ copies/erythrocyte) of the CHIP28 protein on mature erythrocytes.

Where tested, Northern and Western blot analysis confirmed that the antisense cRNA probe hybridization and the antibody staining was due to CHIP28 expression.[103,106] Northern blots were positive (~2.8 kb band) in kidney, lung/airways, spleen and heart, and negative in skeletal muscle, liver and brain. Western blots were positive (28 kDa and 35-60 kDa bands for nonglycosylated and glycosylated CHIP28) in kidney, lung, trachea and heart, and negative in skeletal muscle, liver and brain.

The absence of CHIP28 water channels in certain fluid transporting tissues (such as small intestine) is notable, as is the absence of CHIP28 water channels at one of the two contralateral plasma membranes in many epithelia (such as the basolateral membrane of choroid plexus). It is likely that other distinct water transporting proteins are present in these membranes.

Water Transporters are Responsible for Fluid Movement in Lung

The high expression of CHIP28 in airways and alveoli, and the identification of a CHIP28 homolog in lung (LAW2[104]) suggested that water movement between blood and airway may be mediated by water channels. To investigate this possibility, an in vivo sheep model was used to measure blood-to-airspace osmotic water transport.[67] The airspace was infused with hypertonic saline (900 mOsm) containing a membrane-impermeant marker (³H-albumin), and the pulmonary circulation was perfused with an isosmotic solution. The time course of lung osmotic water movement was determined from the osmolalities and ³H-albumin concentrations in withdrawn samples. The contralateral lung was used to examine whether $HgCl_2$ inhibited transport. It was found that osmotic equilibration occurred rapidly (0.85 min) in control lung, and was remarkably slowed (2.7 min) in the presence of $HgCl_2$. The effect of $HgCl_2$ was fully reversed by mercaptoethanol. These results implicate an important physiological role of lung water channels. Lung water channels may be essential to maintain airspace fluid content during evaporative losses which accompany normal respiration. It was proposed that a major role of type I epithelial cells in alveolus is to transport water.

Localization of MIP26 Family Proteins in Mammalian Tissues

The majority of MIP26 family proteins are derived from plants, bacteria and yeast (see Chapter 7) and are absent in mammalian tissues. The MIP26 protein is very abundant in lens fibers and forms regular orthogonal arrays on freeze fracture electron microscopy.[57] Based on the observation of virtually identical orthogonal arrays in basolateral membrane of kidney collecting ducts,[244] the hypothesis that MIP26 was a basolateral kidney water channel was tested. Northern blotting and antiMIP26-antibody immunostaining studies showed no MIP26 expression in kidney, and oocyte expression and reconstitution studies showed that MIP26 was not a water channel. It was suggested that the unidentified orthogonal arrays in the basolateral membrane of collecting duct principal cells may represent a new member of the MIP26 family.

As described above, the WCH-CD water channels is expressed only in collecting ducts. New members of the MIP26 family recently cloned from rat cDNA libraries (see Chapter 7) appear to localize to kidney collecting ducts (WCH3[146]), kidney and intestine[116] and lung (LAW2[104]). It is anticipated that additional channel-forming proteins in the MIP26 family will be identified and characterized from mammalian tissues in the near future.

Nonwater Transporting Roles of Water Channels

The information presented above establishes an important role for water channels in osmotically-driven transepithelial fluid movement. A role for functional intracellular water channels is less clear; the localization of water channels to subcellular vesicles in water permeable cells may be a consequence of routing to and from plasma membranes as part of a regulated or constitutive trafficking mechanism. A role for rapid swelling of organelles in the membrane fusion has been proposed but is unlikely. The stably transfected CHO cells expressing high levels of functional CHIP28 water channels in plasma membrane and intracellular vesicles were no different from control cells in cell growth and viability, and in kinetics of vesicular trafficking.[147]

The role(s) of water channels in erythrocytes remains uncertain. The need for high water permeability for erythrocyte volume regulation is unlikely, although it is has been speculated that water may be hydrostatically expelled from erythrocytes during passage through narrow splenic capillaries. Erythrocyte volume changes considerably during passage through vasa recta in the hypertonic renal interstitium. A role for erythrocyte urea transporters in attenuating erythrocyte shrinkage and rebound swelling was proposed;[149] however the normal physiology of erythrocytes lacking urea transporters[70] suggests that urea transporters are not necessary. High erythrocyte water permeability would, if anything, increase the minimum and maximum erythrocyte volumes during passage through vasa recta. Based on biochemical data showing high affinity of CHIP28 to lipids, it is proposed that the abundant and hydrophobic CHIP28 water channels may play a structural role in the maintenance of cell membrane integrity during osmotic and mechanical stress. The high density of CHIP28 may organize membrane lipids and proteins, and/or provide attachment points for the erythrocyte membrane skeleton. It would be interesting to study erythrocyte physiology in a transgenic CHIP28 knock-out animal model.

Water channels may play completely unrelated roles in cell growth and differentiation, wound repair and neoplasia. Lanahan et al[136] used a subtractive approach to clone a series of delayed early response genes from growth factor-stimulated fibroblasts. Surprisingly, one of the 15 genes they cloned (DER2) was identical in sequence to rat CHIP28.[46,295] DER2 mRNA was not normally expressed, but was observed on Northern blots at 4-36 hours after growth factor addition. DER2 was also expressed in regenerating liver. It would be interesting to examine the growth characteristics of fibroblasts transfected with an antisense CHIP28 DNA, and to examine whether CHIP28 is expressed in other nonepithelial cells during wound healing and rapid cell proliferation.

REFERENCES

1. Alcayaga C, Cecchi X, Alvarez O, Latorre R. Streaming potential measurements in Ca^{2+}-activated K$^+$ channels from skeletal and smooth muscle. Coupling of ion and water fluxes. Biophys J 1989; 55: 367-71.
2. Al-Zahid B, Schafer JA, Troutman SL, Andreoli TE. Effect of antidiuretic hormone on water and solute permeation, the activation energies of these processes in mammalian cortical collecting tubules. J Membr Biol 1977; 31: 103-129.
3. Bacic G, Srejic R, Ratkovic S. Water transport through membranes: a review of NMR studies of model and biological systems. Studia Biophysica 1990; 138: 95-104.
4. Bae HR, Verkman AS. Protein kinase A regulates chloride conductance in endocytic vesicles from proximal tubule. Nature 1990; 348: 635-637.
5. Baker ME, Saier MH. A common ancestor for bovine lens fiber major intrinsic protein, soybean nodulin-26 protein, *E. coli* glycerol facilitator. Cell 1990; 60: 185-186.
6. Barry PH, Diamond MJ. Effects of unstirred layers on membrane phenomena. Physiol Rev 1984; 64: 763-872.
7. Bartels H, Miragall F. Orthogonal arrays of particles in the plasma membrane of pneumocytes. J Submicrosc Cytol 1985; 18: 637-646.
8. Basset G, Crone C, Saumon G. Fluid absorption by rat lung in situ. Pathways for sodium entry in the luminal membrane of alveolar epithelium. J Physiol 1987; 384: 325-345.
9. Baumert HG, Fasold H. Cross-linking techniques. Meth Enzymol 1979; 172: 584-609.
10. Benga G, Pop VI, Popescu O, Borza V. On measuring the diffusional water permeability of human red blood cells and ghosts by nuclear magnetic resonance. J Biochem Biophys Meth 1990; 21: 87-102.
11. Benga G, Popescu O, Borza V, Pop VI, Muresan A, Mocsy I, Brain A, Wriggleworth A. Water permeability in human erythrocytes: identification of membrane proteins involved in water transport. J Cell Biol 1986; 41: 252-262.
12. Benga G, Popescu O, Pop VI. Water exchange through erythrocytes membranes: nuclear magnetic resonance studies on the effect of inhibitors and of chemical modification of human membranes. Cell Biol Internatl Rep 1983; 7: 807-818.
13. Bentley PJ. The effect of neurohypophysial extracts on water transfer across the wall of the isolated urinary bladder of the toad *Bufo Marinus*. J Endocrinol 1958; 17: 201-209.
14. Berry CA. Characteristics of water diffusion in the rabbit proximal convoluted tubule. Am J Physiol 1985; 249: F729-F738.
15. Berry CA. Water permeability and pathways in the proximal tubule. Am J Physiol 1983; 245: F279-F294 (abstr).
16. Berry CA, Verkman AS. Osmotic gradient dependence of osmotic water permeability in the rabbit proximal convoluted tubule. J Membr Biol 1988; 105: 33-43.

17. Bicknese S, Zimet D, Park J, Van Hoek AN, Shohet SB, Verkman AS. Detection of water proximity to tryptophan residues in proteins by single photon radioluminescence. Submitted.
18. Bicknese S, Periasamy N, Shohet SB, Verkman AS. Cytoplasmic viscosity near the cell plasma membrane: measurement by evanescent field frequency-domain microfluorimetry. Biophys J 1993; 65: 1272-1282.
19. Bicknese S, Shahrohk Z, Shohet SB, Verkman AS. Single photon radioluminescence. I. Theory and spectroscopic properties. Biophys J 1992; 63: 1256-1266.
20. Bondy C, Chin E, Smith BL, Preston GM, Agre P. Developmental gene expression and tissue distribution of the CHIP28 water-channel protein. Proc Natl Acad Sci 1993; 90: 4500-4504.
21. Bordi C, Amherdt M, Perrelet A. Orthogonal arrays of particles in the gastric parietal cell of the rat: differences between superficial and basal cells in the gland and after pentagastrin or metiamide treatment. Anat Rec 1986; 215: 28-34.
22. Bourguet J, Chevalier J, Hugon JS. Alterations in membrane associated particle distribution during antidiuretic challenge in frog urinary bladder epithelium. Biophys J 1976; 16: 627-639.
23. Brahm J. Diffusional water permeability of human erythrocytes and their ghosts. J Gen Physiol 1982; 79: 791-819.
24. Branden C and Tooze J. "Introduction to protein structure". New York: Garland Publishing, 1991: 3-30.
25. Brena BM, Batista-Vievra F, Ryden L, Porath J. Selective adsorption of immunoglobulins and glycosylated proteins on phenylboronate-agarose. J Chromat 1992; 604: 109-115.
26. Brodsky WA, Rehm WS, Dennis WH, Miller DG. Thermodynamic analysis of the intracellular osmotic gradient hypothesis of active water transport. Science 1955; 121: 302-303.
27. Brown D. Membrane recycling and epithelial cell function. Am J Physiol 1989; 256: F1-F12.
28. Brown D. Structural-functional features of antidiuretic hormone-induced water transport in the collecting duct. Sem Neph 1991; 11: 478-501.
29. Brown D, Grosso A, DeSousa RC. Correlation between water flow and intramembrane particle aggregates in toad epidermis. Am J Physiol 1983; 245: C334-C342.
30. Brown D, Orci L. Vasopressin stimulates formation of coated pits in rat kidney collecting ducts. Nature 1983; 302: 253-255.
31. Brown D, Verbavatz J-M, Valenti G, Lui B, Sabolic I. Localization of the CHIP28 water channel in reabsorptive segments of the rat male reproductive tract. Eur J Cell Biol 1993; 61: 264-273.
32. Brown PA, Feinstein MD, Sha'afi RI. Role of erythrocyte band 3 in water transport. Nature 1975; 254: 523-525.
33. Burg M and Green N. Function of thick ascending limb of Henle's loop. Am J Physiol 1973; 224: 659-668.
34. Calamita G, Valenti G, Svelto M, Bourguet J. Selected polyclonal antibodies and ADH challenge in frog urinary bladder: A label-fracture study. Am J Physiol 1992; 262: F267-F274.
35. Carruthers A, Melchior DL. Studies of the relationship between bilayer physical state. Biochem 1983; 22: 5797-5807.
36. Chang CT, Wu CC, Yang JT. Circular dichroic analysis of protein confomation:

Inclusion of the B-turns. Anal Biochem 1978; 91: 13-31.

37. Chavez RA, Hall ZW. The transmembrane topology of the amino terminus of the alpha subunit of the nicotinic acetylcholine receptor. J Biol Chem 1991; 266: 15532-15538.

38. Chen P-Y, Pearce D, Verkman AS. Membrane water and solute permeability determined quantitatively by self-quenching of an entrapped fluorophore. Biochem 1988; 27: 5713-5719.

39. Chen P-Y, Verkman AS. Nonelectrolyte transport across renal proximal tubule cell membranes measured by tracer efflux and light scattering. Pflugers Arch 1987; 408: 491-496.

40. Chiu SW, Novotny JA, Jabkobson E. The nature of ion and water barrier crossings in a simulated ion channel Biophys J 1993; 64: 98-109.

41. Coleman RA, Wade JB. Role of nonacidic endosomes in recycling of ADH-sensitive water channel. Eur J Cell Biol 1992; 58: 44-56.

42. Corman B, Di Stefano A. Does water drag solutes through the kidney proximal tubule? Pflugers Arch 1983; 397: 35-41.

43. Dani JA, Levitt DG. Water transport and ion-water interaction in the gramicidin channel. Biophys J 1981; 35: 501-508.

44. Da Silva P, Nicolson GL. Membrane intercalated particles in human erythrocyte ghosts: sites of preferred passage of water molecules at low temperature. Proc Natl Acad Sci USA 1973; 70: 1339-1343.

45. De Groot, SR. Thermodyanamics of Irreversible Processes. Amsterdam: North Holland Publishing, 1958.

46. Deen PMT, Dempster JA, Wieringa B, Van Os CH. Isolation of a cDNA for rat CHIP28 water channel: high mRNA expression in kidney cortex and inner medulla. Biochim Biophys Res Comm 1992; 188: 1267-1273.

47. Deen PMT, Weghuis DO, van Kessel AG, Wierigna B, Van Os, CH. The human gene for water channel CHIP28 is localized on chromosome 7p14-15. Cell Gen 1993; in press.

48. Delauney AJ, Cheon CI, Snyder PJ, Verma DP. A nodule-specific sequence encoding a methionine-rich polypeptide, nodulin-21. Plant Molec Biol 1990; 14: 449-51.

49. Dempster JA, Van Hoek AN, de Jong MD, Van Os CH. Glucose transporters do not serve as water channels in renal and intestinal epithelia. Pflugers Arch 1991; 419: 249-255.

50. Dempster JA, Van Hoek AN, Van Os CH. The quest for water channels. News Physiol Sci 1992; 7: 172-176.

51. Deng QS and Johanson CE. Cyclic AMP alteration of chloride transport into choroid plexus-cerebrospinal fluid system. Neurosci Lett 1992: 143: 146-150.

52. Denker BM, Smith BL, Kuhajda FP, Agre P. Identification, purification, partial characterization of a novel Mr 28,000 integral membrane protein from erythrocytes and renal tubules. J Biol Chem 1988; 263: 15634-15642.

53. Dermietzel R. Junctions in the central nervous system of the cat. Gap junctions and membrane-associated orthogonal particle complexes in astrocytic membranes. Cell Tissue Res 1974; 149: 121-135.

54. Dix JA, Ausiello DA, Jung CY, Verkman AS. Target analysis studies of red cell water and urea transport. Biochim Biophys Acta 1985; 821: 243-252.

55. Donaldson P, Kistler J. Reconstitution of channels from preparations enriched in lens gap junction protein MP70. J Membr Biol 1992; 129: 155-65.

56. Donnelly D, Overington JP, Ruffle SV, Nugent JHA, Blundell TL. Modeling

α-helical transmembrane domains: The calculation and use of substitution tables for lipid-facing residues. Protein Science 1993; 2: 55-70.

57. Dunia I, Manenti S, Rousselet A, Benedetti EI. Electron microscopic observations of reconstituted proteoliposomes with the purified major intrinsic membrane protein of eye lens fiber. J Cell Biol 1987; 195: 1679-1689.

58. Echevarria M, Frindt G, Preston GM, Milovanovic S, Agre P, Fischbarg J, Windhager EE. Expression of multiple water channel activities in Xenopus oocytes injected with messenger RNA from rat kidney. J Gen Physiol 1993; 101: 827-841.

59. Echevarria M, Verkman AS. Optical measurement of osmotic water transport in cultured cells: evaluation of the role of glucose transporters. J Gen Physiol 1992; 99: 573-589.

60. Farinas J, Van Hoek AN, Shi L-B, Erickson C, Verkman AS. Nonpolar environment of tryptophans in erythrocyte water channel CHIP28 determined by fluorescence quenching. Biochem 1993; in press.

61. Fettiplace R, Haydon DA. Water permeability of lipid membranes. Physiol Rev 1980; 60: 510-550.

62. Finkelstein A. Water and nonelectrolyte permeability of lipid bilayer membranes. J Gen Physiol 1976; 68: 127-135.

63. Finkelstein A. Water Movement through Lipid Bilayers, Pores, Plasma Membranes: Theory and Reality. New York: Wiley & Sons 1987.

64. Fischbarg J, Kunyan K, Hirsch J, Lecuona S, Rogozinski L, Silverstein S, Loike J. Evidence that the glucose transporter serves as a water channel in J774 macrophages. Proc Natl Acad Sci USA 1989; 86: 8397-8401.

65. Fischbarg J, Kunyan K, Vera JC, Arant S, Silverstein S, Loike J, Rosen OM. Glucose transporters serve as water channels. Proc Natl Acad Sci USA 1990; 87: 3244-3247.

66. Flamion B, Spring KR. Water permeability of apical and basolateral cell membranes of rat inner medullary collecting duct. Am J Physiol 1990; 259: F986-F999.

67. Folkesson HG, Hasegawa H, Kheradmand F, Matthay MA, Verkman AS. Transcellular water transport in lung alveolar epithelium through mercurial-sensitive water channels. Submitted.

68. Frigeri A, Lema F, Monier F, Bourguet J. Selective labeling of apical membrane proteins of frog urinary bladder. Antibodies to frog bladder cross react with principal cells of collecting duct of rabbit kidney. Molec Biol Cell 1992; 2: 1667.

69. Frigeri A, Ma T, Verkman AS. Molecular cloning of a family of channel-forming proteins homologous to CHIP28 from rat and human kidney. JASN 1993; 4: 852(abstr).

70. Frolich O, Macey RI, Edwards-Moulds J, Gargus JJ, Gunn RB. Urea transport deficiency in Jk(a-b-) erythocytes. Am J Physiol 1991; 260: C778-83.

71. Fujiki Y, Hubbard AL, Fowler S, Lazarow PB. Isolation of intracellular membranes by means of sodium carbonate treatment: Application to endoplasmic reticulum. J Cell Biol 1982; 93: 97-102.

72. Fushimi K, Dix JA, Verkman AS. Cell membrane fluidity in the intact kidney proximal tubule measured by fluorescence anisotropy imaging. Biophys J 1990; 57: 241-254.

73. Fushimi K, Uchida S, Hara Y, Hirata Y, Marumo F, Sasaki S. Cloning and expression of apical membrane water channel of rat kidney collecting tubule. Nature 1993; 361: 549-552.

74. Fushimi K, Verkman AS. Low viscosity in the aqueous domain of cell cytoplasm measured by picosecond polarization microscopy. J Cell Biol 1991; 112: 719-725.

75. Fushimi K, Verkman AS. Relationship between plasma membrane fluidity and vasopressin-sensitive water permeability in kidney collecting tubule. Am J Physiol 1991; 260: C1-C8.
76. Garavito RM. Crystallizing membrane proteins: experiments on different systems. In: Michel H, ed. Crystallization of Membrane Proteins, Boca Raton: CRC Press, 1991: 89-106.
77. Giocondi MC, Friedlander G, Le Grimellec C. ADH modulates plasma membrane lipid order of living MDCK cells via a cAMP-dependent process. Am J Physiol 1990; 259: F95-103.
78. Giocondi MC, Le Grimellec C. Water permeation in Madin-Darby canine kidney cells is modulated by membrane fluidity. Biochim Biophys Acta 1991; 1064: 315-320.
79. Girsch SJ, Perrachia C. Calmodulin interacts with a C-terminus peptide from the lens membrane protein MIP26. Cur Eye Res 1991; 10: 839-49.
80. Goldstein DA, Solomon AK. Determination of equivalent pore radius of human red cells by osmotic pressure measurements. J Gen Physiol 1960; 44: 1-17.
81. Gonzales E, Capri-Medina P, Whittembury M. Cell osmotic water permeability of isolated rabbit tubule. Am J Physiol 1982; 242: F321-F330.
82. Gorin MB, Yancey SB, Cline J, Revel J-P, Horwitz J. The major intrinsic protein (MIP) of the bovine lens fiber membrane: characterization and structure based on cDNA cloning. Cell 1984: 39: 49-59.
83. Grantham JJ. Vasopressin: effect on deformability of urinary surface of collecting duct cells. Science 1968; 168: 1093-1095.
84. Grantham JJ, Berg MB. Effect of vasoressin and cyclic AMP on permeability of isolated collecting tubules. Am J Physiol 1966; 211: 255-259.
85. Green R, Giebisch G. Luminal hypotonicity: a driving force for fluid absorption from the proximal tubule. Am J Physiol 1984; 246: F167-174.
86. Green R, Giebisch G. Reflection coefficients and water permeability in rat proximal tubule. Am J Physiol 1989; 257: F658-F668.
87. Gronowicz G, Masur SK, Holtzman E. Quantitative analysis of exocytosis and endocytosis in the hydroosmotic response of toad urinary bladder. J Membr Biol 1980; 52: 221-235.
88. Grossman EB, Harris HW, Star RA, Zeidel ML. Water and nonelectrolyte permeabilities of apical membranes of toad urinary bladder granular cells. Am J Physiol 1992; 262: C1109-C1118.
89. Guerrero FD, Jones JT, Mullet JE. Turgor-responsive gene transcription and RNA levels increase rapidly when pea shoots are wilted. Sequence and expression of three inducible genes. Plant Molec Biol 1990; 15: 11-26.
90. Haas M, McBrayer DG, Yankaskas JR. Dual mechanisms for Na-K-Cl transport regulation in airway epithelial cells. Am J Physiol 1993; 264: C189-C200.
91. Handler JS. Antidiuretic hormone moves membranes. Am J Physiol 1988; 255: F375-F382.
92. Harmanci MC, Stern P, Kachadorian WA, Valtin H, DiScala VA. Vasopressin and collecting duct intramembranous particle clusters: a dose-response relationship. Am J Physiol 1983; 239: F560-F564.
93. Harris HW, Paredes A, Hosselet C, Agre P, Preston G, Strange K, Siner J. Cloning of a CHIP28 homolog localized to water channel containing endosomes (WCV) in toad urinary bladder (TB). JASN, 1993; 4: 853(abstr).
94. Harris HW, Botelho B, Zeidel ML, Strange K. Cytoplasmic dilution induces antidiuretic hormone water channel retrieval in toad urinary bladder. Am J

Physiol 1992; 263: F163-F170.
95. Harris HW, Handler JS, Blumenthal R. Apical membrane vesicles of ADH-stimulated toad bladder are highly water permeable. Am J Physiol 1990; 258: F237-F243.
96. Harris HW, Hosselet C, Guaywoodford L, Zeidel ML. Purification and partial characterization of candidate antidiuretic hormone water channel proteins of M(r) 55,000 and 53,000 from toad urinary bladder. J Biol Chem 1992; 267: 22115-22121.
97. Harris HW, Kikeri D, Janoshazi A, Solomon AK. High proton flux through membranes containing antidiuretic hormone water channels. Am J Physiol 1990; 259: F366-F371.
98. Harris HW, Strange K, Zeidel M. Current understanding of the cellular biology and molecular structure of the antidiuretic hormone-stimulated water transport pathway. J Clin Invest 1991; 91: 1-8.
99. Harris HW, Wade JB, Handler JS. Identification of specific apical membrane polypeptides associated with the antidiuretic hormone-elicited water permeability increase in the toad urinary bladder. Proc Natl Acad Sci USA 1988; 85: 1942-46.
100. Harris HW, Wade JB, Handler JS. Transepithelial water flow regulates apical membrane retrieval in antidiuretic hormone-stimulated toad urinary bladder. J Clin Invest 1986; 78: 703-712.
101. Harris HW, Zeidel ML, Hosselet C. Quantitation and topography of membrane proteins in highly water-permeable vesicles from ADH-stimulated toad bladder. Am J Physiol 1991; 261: C143-C153.
102. Harvey B, Lacoste I, Ehrenfeld J. Common channels for water and protons at apical and basolateral cell membranes of frog skin and urinary bladder epithelia. Effects of oxytocin, heavy metals, inhibitors of H(+)-adenosine triphosphatase. J Gen Physiol 1991; 97: 749-76.
103. Hasegawa H, Lian SC, Finkbeiner WE, Verkman AS. Extrarenal tissue distribution of CHIP28 water channels by in situ hybridization and antibody staining. Am J Physiol 1993; in press.
104. Hasegawa H, Ma T, Matthay M, Finkbeiner W, Verkman AS. Identification and cloning of extrarenal water channels. JASN 1993; 4: 854 (abstr).
105. Hasegawa H, Skach W, Baker O, Calayag MC, Lingappa VR, Verkman AS. A multifunctional aqueous channel formed by CFTR. Science 1992; 258: 1477-1479.
106. Hasegawa H, Zhang R, Dohrman A, Verkman AS. Tissue-specific distribution of mRNA encoding rat kidney water channel CHIP28k by in situ hybridization. Am J Physiol 1993; 264: C237-C245.
107. Hasegawa H, Verkman AS. Functional expression of cAMP-dependent and independent urea transporters in *Xenopus* oocytes. Am J Physiol 1993; 265: C514-C520.
108. Hays RM. Alteration of luminal membrane structure by antidiuretic hormone. Am J Physiol 1983; 245: C289-C296.
109. Holz R, Finkelstein A. The water and nonelectrolyte permeability induced in thin lipid membranes by the polyene antibiotics nystatin and amphotericin B. J Gen Physiol 1970; 56: 125-145.
110. House CR. Water Transport in Cells and Tissues. Baltimore: Williams & Wilkins, 1974.
111. Hugon JS, Ibarra C, Valenti G, Bourguet J. Microtubules and actin microfila-

ments in the amphibian bladder granular cells. Biol Cell 1989; 66: 77-84.
112. Ibarra C, Ripoche P, Parisi M, Bourguet J. Effects of PCMBS on the water and small solute permeabilities in frog urinary bladder. J Membr Biol 1990; 116: 57-64.
113. Ibarra C, Ripoche P, Bourguet J. Effect of mercurial compounds on net water transport and intramembrane particle aggregates in ADH-treated frog urinary bladder. J Membr Biol 1989; 110: 115-126.
114. Illsley NP, Verkman AS. Serial barriers to water permeability in human placental vesicles. J Membr Biol 1986; 94: 267-278.
115. Imai M, Yoshitomi K. Heterogeneity of the descending thin limb of Henle's loop. Kid. Internatl 1990; 38: 687-694.
116. Ishibashi, K., Sasaki S, Fushimi K, Uchida S, Muwahara M, Marumo F. Molecular cloning and expressing of new water channel (WCH-3) expressed selectively in kidney and GI tract. JASN 1993; 4: 855 (abstr).
117. Jacobson HR, Kokko JP, Seldin DW, Holmberg C. Lack of solvent drag of NaCl and NaHCO$_3$ in rabbit proximal tubules. Am J Physiol 1982; 243: F342-F348.
118. Jähnig F. Structure prediction for membrane proteins. In: Prediction of Protein Structure and the Principles of Protein of Conformation, Fasman GD, ed. Plenum Press, NY and London 1989; 707-717.
119. Jansson T, Illsley NP. Osmotic water permeabilities of human placental microvillous and basal membranes. J Membr Biol 1993; 132: 147-145.
120. Johnson KD, Hofte H, Chrispeels MJ. An intrinsic tonoplast protein of protein storage vacuoles in seeds is structurally related to a bacterial solute transporter (GlpF). Plant Cell 1990; 2: 525-532.
121. Kabsch W, Sander C. Dictionary of protein secondary structure: pattern recognition of hydrogen-bonded and geometrical factors. Biopolymers 1983; 22: 2577-2639.
122. Kao HP, Abney JR, Verkman AS. Determinants of the translational diffusion of a small solute in cell cytoplasm. J Cell Biol 1993; 120: 175-184.
123. Kao HP, Verkman AS. Tracking of single fluorescent particles in three dimensions by a point-spread function-based autofocus method. Biophys J 1992; 64: A219.
124. Katchalsky A, Curran PF. Nonequilibrium thermodynamics in biophysics. Cambridge: Harvard Univ Press.
125. Kavenau JL. Water and Solute-Water Interactions. San Francisco: Holden-Day Publishers, 1964.
126. Kedem O, Katchalsky A. A physical interpretation of the phenomenological coefficients of membrane permeability. J Gen Physiol 1961; 45: 143-179.
127. Kedem O, Katchalsky A. Thermodynamic analysis of the permeability of biological membranes to nonelectrolytes. Biochim Biophys Acta 1958; 27: 229-246.
128. Kempner ES, Verkman AS. Direct effects of ionizing radiation unique to macromolecules. Radiat Phys Chem 1988; 32: 341-347.
129. Komissarchik YY, Snigirevskaya ES. Participation of intracellular membranes in the formation of high permeable domains in plasma membranes of epithelial cells during vasopressin stimulation of water transport. Tsitologiya (Russian) 1991; 33: 11-135.
130. Krejs GJ, Fordtran JS. Physiology and pathophysiology of ion and water movement in the human intestine. In: Sleisenge MJ and Fordtran JS, eds. Gastrointestinal Disease: Pathophysiology, Diagnosis and Management. Phila-

delphia: Sanders, 1978; 2: 297-312.

131. Kuwahara M, Berry CA, Verkman AS. Rapid development of vasopressin-induced hydroosmosis in kidney collecting tubules measured by a new fluorescence technique. Biophys J 1988; 54: 595-602.

132. Kuwahara M, Sasaki S, Fushimi K, Marumo F. Protein kinase A regulates water channel of kidney collecting duct (WCH-CD) expressed in *Xenopus* oocytes. JASN 1993; 1993; 4: 855 (abstr).

133. Kuwahara M, Shi L-B, Marumo F, Verkman AS. Transcellular water flow modulates water channel exocytosis and endocytosis in kidney collecting tubule. J Clin Invest 1991; 88: 423-429.

134. Kuwahara M, Verkman AS. Direct fluorescence measurement of diffusional water permeability in the vasopressin-sensitive kidney collecting tubule. Biophys J 1988; 54: 587-593.

135. Kuwahara M, Verkman AS. Pre-steady-state analysis of the regulation of water permeability in the kidney collecting tubule. J Membr Biol 1989; 110: 57-65.

136. Lanahan A, Williams JB, Landers LK, Nathans D. Growth factor-induced delayed early response genes. Mol Cell Biol 1992; 12: 3919-3929.

137. Lawaczeck R. Water permeability through biological membranes by isotopic effects of fluorescence and light scattering. Biophys J 1984; 45: 491-494.

138. Leaf A, Hays RM. Permeability of the isolated toad bladder to solutes and its modification by vasopressin. J Gen Physiol 1962; 45: 921-932.

139. Lencer WI, Brown D, Ausiello DA, Verkman AS. Endocytosis of water channels in rat kidney: cell specificity and correlation with in vivo antidiuretic states. Am J Physiol 1990; 259: 920-932.

140. Lencer WI, Verkman AS, Ausiello DA, Arnaout A, Brown D. Endocytic vesicles from renal papilla which retrieve the vasopressin-sensitive water channel do not contain an H^+ ATPase. J Cell Biol 1990; 111: 379-389.

141. Lencer WI, Weyer P, Verkman AS, Ausiello DA, Brown D. FITC-dextran as a probe for endosome function and localization in kidney. Am J Physiol 1990; 258: C309-C317.

142. Levine S, Jacoby M, Finkelstein A. The water permeability of toad urinary bladder. II. The value of P_f/P_d for antidiuretic hormone-induced water permeation pathway. J Gen Physiol 1984; 83: 543-561.

143. Levitt DG, Mlekoday HJ. Reflection coefficient and permeability of urea and ethyleneglycol in the human red cell membrane. J Gen Physiol 1983; 81: 239-254.

144. Lipp J, Flint N, Haeuptle MT, Dobberstein B. Structural requirements for membrane assembly of proteins spanning the membrane several times. J Cell Biol 1989; 109: 2013-2022.

145. Lukacovic MF, Verkman AS, Dix JA, Solomon AK. Specific interaction of a water transport inhibitor with band 3 in red blood cell membranes. Biochim Biophys Acta 1984; 778: 253-259.

146. Ma T, Frigeri A, Skach W, Verkman AS. Cloning of a novel rat kidney cDNA homologous to CHIP28 and WCH-CD water channels. Submitted.

147. Ma T, Frigeri A, Tsai S-T, Verbavatz JM, Verkman AS. Localization and functional analysis of CHIP28k water channels in stably transfected CHO cells. J Biol Chem 1993; 256: 22756-22764.

148. Ma T, Hasegawa H, Skach W, Frigeri A, Verkman AS. Expression, functional analysis and in situ hybridization of a cloned rat kidney collecting duct water channel. Am J Physiol 1993; in press.

149. Macey RI. Transport of water and urea in red blood cells. Am J Physiol 1984;

246: C195-C203.
150. Macey RI, Farmer REL. Inhibition of water and solute permeability in human red cells. Biochim Biophys Acta 1970; 211: 104-106.
151. Martial S, Ripoche P. An ultrarapid filtration method adapted to the measurements of water and solute permeability of synthetic and biological vesicles. Anal Biochem 1991; 197: C296-304.
152. Masters BR, Yguerabide J, Fanestil DD. Microviscosity of mucosal cellular membranes in toad urinary bladder: relation to antidiuretic hormone action on water permeability. J Membr Biol 1978; 40: 179-190.
153. Masur SK, Cooper S, Rubin MS. Effect of an osmotic gradient on antidiuretic hormone-induced endocytosis in the hydroosmosis in the toad urinary bladder. Am J Physiol 1984; 247: F370-F379.
154. Masur SK, Massardo S. ADH and phorbol ester increase immunolabeling of the toad bladder apical membrane by antibodies made to granules. J Membr Biol 1987; 96: 193-198.
155. Maurel C, Reizer J, Schoeder JI, Chrispeels MJ. The vacuolar membrane protein gamma-TIP creates water specific channels in Xenopus oocytes. Embo J 1993; 12: 2241-7.
156. McKie AT, Goecke IA, Naftalin RJ. Comparison of fluid absorption of bovine and ovine descending colon in vitro. Am J Physiol 1991; 261: F433-F442.
157. Mellman I, Fuchs R, Helenius A. Acidification of the endocytic and exocytic pathways. Annu Rev Biochem 1986; 55: 663-700.
158. Melton DA, Krieg PA, Bebagliati MR, Maniatic T, Zinn K, Green MR. Efficient in vitro synthesis of biologically active RNA and RNA hybridization probes from plasmids containing a bacterial SP6 promotor. Nucl Acids Res 1984; 12: 7035-7056.
159. Meyer MM, Verkman AS. Evidence for water channels in proximal tubule cell membranes. J Membr Biol 1987; 96: 107-119.
160. Meyer MM, Verkman AS. Human platelet osmotic water and non-electrolyte transport. Am J Physiol 1986; 250: C549-C557.
161. Miao GH, Hong Z, Verma DP. Topology and phosphorylation of soybean nodulin-26, an intrinsic protein of the peribacteroid membrane. J Cell Biol 1992; 118: 481-90.
162. Mlekoday HJ, Moore R, Levitt DG. Osmotic water permeability of the human red cell: dependence on direction of water flow and cell volume. J Gen Physiol 1983; 81: 213-220.
163. Moura TR, Macey RI, Chien DY, Daran D, Santos H. Thermodynamics of all-or-none water channels closure in red cells. J Membr Biol 1984; 81: 105-111.
164. Muallem S, Zhang R, Loessberg PA, Star RA. Simultaneous recording of cell volume changes intracellular pH or Ca^{2+} concentration in single osteosarcoma cells UMR-106-01. J Biol Chem 1992; 267: 17658-17664.
165. Muller J, Kachadorian A, DiScala VA. Evidence that ADH-stimulated intramembrane particle aggregates are transferred from cytoplasmic to luminal membranes in toad bladder epithelial cells. J Cell Biol 1980; 85: 83-93.
166. Muramatsu S, Mizuno T. Nucleotide sequence of the region encompassing the glpKF operon and its upstream region containing a bent DNA sequence of Eschirichia coll. Nucl Acids Res 1989; 17: 4378.
167. Naftalin RJ, Pedley KC. Video enhanced imaging of the fluorescent Na+ probe SBFI indicates that colonic crypts absorb fluid by generating a hypertonic interstitial fluid. FEBS Lett 1990; 260: 187-194.

168. Narula P, Xu M, Kuang KY, Akiyama R, Fischbarg J. Fluid transport across cultured bovine corneal endothelial cell monolayers. Am J Physiol 1992; 262: C98-C103.
169. Nielsen S, Christensen EI, DiGiovanni SR, Harris HW, Knepper MA. Immunolocalization of vasopressin-regulated water channel in rat kidney. JASN 1993; 4: 857 (abstr).
170. Nielsen S, Smith BL, Christensen EI, Knepper MA, Agre P. CHIP28 water channels are localized in constitutively water-permeable segments of the nephron. J Cell Biol 1993; 120: 371-383.
171. Ojcius DM, Solomon AK. Sites of p-chloromercuribenzene sulfonate inhibition of red cell urea and water transport. Biochim Biophys Acta 1988; 942: 73-82.
172. Pao GM, Wu L-F, Johnson KD, Hofte H, Chrispeels MJ, Sweet G, Sandal NN, Saier MH. Evolution of the MIP family of integral membrane transport of proteins. Molec Microbiol 1991; 5: 33-37.
173. Park K, Perczel A, Fasman G. Differentiation between transmembrane helices and peripheral helices by the deconvolution of circular dichroism spectra of membrane proteins. Protein Science, 1992; 1: 1032-1049.
174. Pearce D, Verkman AS. NaCl reflection coefficients in proximal tubule apical and basolateral vesicles: measurement by induced osmosis and solvent drag. Biophys J 1989; 55: 1251-1259.
175. Perara E, Rothman RE, Lingappa VR. Uncoupling translocation from translation: implications for transport of proteins across membranes. Science 1986; 232: 348-352.
176. Periasamy N, Armijo M, Verkman AS. Picosecond rotation of small polar fluorophores in the cytosol of sea urchin eggs. Biochem 1991; 30: 11836-11841.
177. Periasamy N, Kao HP, Fushimi K, Verkman AS. Organic osmolytes increase cytoplasmic microviscosity in kidney cells. Am J Physiol 1992; 263: C901-C907.
178. Persson BE, Spring KR. Gallbladder epithelial cell hydraulic water permeability and volume regulation. J Gen Physiol 1982; 79: 481-505.
179. Pfeffer W. "Osmotische Untersuchungen" Leipzig, Verlag von Wilhelm Engelmann. 1877.
180. Pietras RJ, Wright EM. Nonelectrolyte probes of membrane structure in ADH-treated toad urinary bladder. Nature 1974; 247: 222-224.
181. Pratz JP, Ripoche P, Corman B. Evidence for proteic water pathways in the luminal membrane of kidney proximal tubule. Biochim Biophys Acta 1986; 856: 259-266.
182. Pratz JP, Ripoche P, Corman B. Osmotic water permeability and solute reflection coefficients of rat kidney brush-border membrane vesicles. Biochim Biophys Acta 1986; 861: 395-397.
183. Preisig PA, Berry CA. Evidence for transcellular osmotic water flow in rat proximal tubules. Am J Physiol 1985; 249: F124-F131.
184. Preston GM, Agre P. Isolation of the the cDNA for erythrocyte integral membrane protein of 28-kilodaltons—member of an ancient channel family. Proc Natl Acad Sci USA 1991; 88: 11110-11114.
185. Preston BM, Carroll TP, Guggino WB, Agre P. Appearance of water channels in Xenopus oocytes expressing red cell CHIP28 protein. Science 1992; 256: 385-387.
186. Preston GM, Jung JS, Guggino WB, Agre P. The mercury-sensitive residue at cysteine 189 in the CHIP28 water channel. J Biol Chem 1993; 268: 17-20.
187. Priver NA, Rabon EC, Zeidel ML. Apical membrane of the gastric parietal

cell—water, proton, nonelectrolyte permeabilities. Biochem 1993; 32: 2459-2468.
188. Rao Y, Jan LY, Jan YN. Similarity of the product the Drosophila neurogenic gene big brain to transmembrane channel proteins. Nature 1990; 345: 163-167.
189. Reif MC, Troutman SL, Schafer JA. Sustained response to vasopressin in isolated rat cortical collecting tubule. Kid Internatl 1984; 26: 725-732.
190. Rodman JS, Seidman L, Farquhar MG. The membrane composition of coated pits, microvilli, endosomes and lysosomes in distinctive in rat the rat kidney proximal tubule cell. J Cell Biol 1986; 102: 77-87.
191. Rothman RE, Andrews DW, Calayag MC, Lingappa VR. Construction of defined polytopic integral transmembrane proteins. J Biol Chem 1988; 263: 10470-10480.
192. Sabolic I, Shi L-B, Brown D, Ausiello DA, Verkman AS. Proteinases inhibit H-ATPase and Na-H exchange but not water transport in apical and endosomal vesicles from rat proximal tubule. Biochim Biophys Acta 1992; 1103: 137-147.
193. Sabolic I, Valenti G, Verbavatz JM, Van Hoek AN, Verkman AS, Ausiello DA, Brown D. Localization of the CHIP28 water channel in rat kidney. Am J Physiol 1992; 263: C1225-C1233.
194. Sabolic I, Wuarin F, Shi L-B, Verkman AS, Ausiello DA, Gluck S, Brown D. Apical endosomes from collecting duct principal cells lack subunits of the proton pumping ATPase. J Cell Biol 1992; 19: 111-122.
195. Sandal NN, Marcker KA. Soybean nodulin 26 is homologous to the major intrinsic protein of the bovine lens fiber membrane. Nucl Acids Res 1988; 16: 9347-9348.
196. Sarver RW, Krueger W. Protein secondary structure from Fourier transform infrared spectroscopy: a data base analysis. Anal Biochem 1990; 194: 89-100.
197. Sasaki S, Saito H, Saito F, Fushimi K, Uchida S, Rai Y, Ikeuhi T, Inui K, Marumo F. Cloning, expression and chromosomal mapping of human collecting duct water channel. JASN 1993; 4: 858 (abstr).
198. Schafer JA, Troutman SL, Andreoli TE. Osmosis in cortical collecting tubules: ADH-independent osmotic flow rectification. J Gen Physiol 1974; 64: 228-240.
199. Shahrohk Z, Bicknese S, Shohet SB, Verkman AS. Single photon radioluminescence. II. Experimental detection and biological applications. Biophys J 1992; 63: 1267-1279.
200. Shen L, Shrager P, Girsch SJ, Donaldson PJ, Peracchia C. Channel reconstituion in liposomes and planar bilayers with HPLC-purified MIP26 of bovine lens. J Membr Biol 1991; 124: 21-32.
201. Shi L-B, Brown D, Verkman AS. Water, urea and proton transport properties of endosomes containing the vasopressin-sensitive water channel from toad bladder. J Gen Physiol 1990; 95: 941-60.
202. Shi L-B, Fushimi K, Bae HR, Verkman AS. Heterogeneity in acidification measured in individual endocytic vesicles isolated from kidney proximal tubule. Biophys J 1991; 59: 1208-1217.
203. Shi L-B, Fushimi K, Verkman AS. Solvent drag measurement of transcellular and basolateral membrane NaCl reflection coefficient in proximal tubule. J Gen Physiol 1991; 98: 379-398.
204. Shi-B and Verkman AS. Very high-water permeability in vasopressin-dependent endocytic vesicles in toad urinary bladder. J Gen Physiol 1989; 94: 1101-1115.
205. Shi L-B, Wang Y-X, Verkman AS. Regulation of the formation and water permeability of endosomes from toad bladder granular cells. J Gen Physiol 1990; 96: 789-808.

206. Shiels A, Ket NA, McHale M, Bangham JA. Homology of MIP26 to NOD26. Nucl Acid Res 1998; 16: 9348.
207. Sigel R. Use of Xenopus oocytes for the functional expression of plasma membrane proteins. J Membr Biol 1990; 117: 201-221.
208. Skach WR, Calayag MC, Lingappa VR. Evidence for an alternate model of human P-glycoprotein structure and biogenesis. J Biol Chem 1993; 269: 6903-6908.
209. Skach WR, Lingappa VR. In: Loh P, ed. Intracellular Trafficking of Pre(pro-) Proteins Across RER Membranes. Boca Raton: CRC Press 1993: 19-77.
210. Skach W, Shi L-B, Calayag MC, Lingappa VR, Verkman AS. Membrane topology of the human CHIP28 water channel. J Cell Biol 1993.
211. Smith BL, Agre P. Erythrocyte M_r 28,000 transmembrane protein exists as a multisubunit oligomer similar to channel proteins. J Biol Chem 1991; 266: 6407-6415.
212. Solomon AK. Characterization of biological membranes by equivalent pores. J Gen Physiol 1968; 51: 335s-364s.
213. Solomon AK, Chasan B, Dix JA, Lukacovic MF, Toon MR, Verkman AS. The aqueous pore in the red cell membrane: band 3 as a channel for anions, cations, nonelectrolytes and water. Proc NY Acad Sci 1984; 414: 79-124.
214. Staros JV, Anjaneyulu PS. Membrane-impermeant cross-linking reagents. Meth Enzymol 1979; 172: 609-629.
215. Strange K, Spring KR. Cell membrane water permeability of rabbit cortical collecting duct. J Membr Biol 1987; 96: 27-43.
216. Strange K, Spring KR. Methods of imaging renal tubule cells. Kid Internatl 1986; 30: 192-200.
217. Strange K, Willingham MC, Handler JS, Harris HW. Apical membrane endocytosis via coated pits is stimulated by removal of antidiuretic hormone from isolated, perfused rabbit cortical collecting tubule. J Membr Biol 1988; 103: 17-28.
218. Swamy MC, Abraham EC. Glycation of lens MIP26 affects the permeability in reconstituted liposomes. Biochem Biophys Res Comm 1992; 186: 632-638.
219. Sweet G, Gandor C, Voegele R, Wittekindt N, Beuerle J, Truniger V, Lin ECC, Boos W. Glycerol facilitator of *Escherichia coli*: cloning of GlpF and identification of the GlpF product. J Bacteriol 1990; 172: 424-430.
220. Terwilliger TC, Solomon AK. Osmotic water permeability of human red cells. J Gen Physiol 1981; 77: 549-570.
221. Thevenin BJM, Periasamy N, Shohet SB, Verkman AS. Segmental dynamics of the cytoplasmic domain of erythrocyte band 3 determined by time-resolved fluorescence anisotropy: sensitivity to pH and ligand binding. Submitted.
222. Tsai S-T, Zhang R, Verkman AS. High channel-mediated water permeability in rabbit erythrocytes: characterization in native cells and Xenopus oocytes. Biochem 1991; 30: 2087-2092.
223. Tytgat J, Hess P. Evidence for cooperative interactions in potassium channel gating. Nature 1992; 359: 420-423.
224. Valenti G, Verbavatz JM, Sabolic I, Ausiello DA, Verkman AS, Brown D. CHIP28/MIP26-related protein in basolateral membranes of kidney principal cells and gastric parietal cells. Submitted.
225. Valenti G, Calamita G, Svelto M. Polyclonal antibodes in study of ADH-induced water channels in frog urinary bladder. Am J Physiol 1991; 261: F437-F442.

226. Valenti G, Casavola V, Svelto M. Isolation of frog urinary bladder plasma membranes with polycation coated beads. Biol Cell 1989; 66: 85-89.

227. Valenti G, Guerra L, Casavola V, Svelto M. Fluorescence labeling of proteins related to ADH-induced change in frog bladder luminal membrane. Biol Cell 1989; 67: 115-121.

228. Van der Goot F, Corman B. Axial heterogeneity of apical water permeability along rabbit kidney proximal tubule. Am J Physiol 1991; 260: 186-191.

229. Van der Goot F, Corman B, Ripoche P. Evidence for permanent water channels in the basolateral membrane of an ADH-sensitive epithelium. J Membr Biol 1991; 120: 59-65.

230. Van der Goot F, Podevin RA, Corman B. Water permeabilities and salt reflection coefficients of luminal, basolateral and intracellular membrane vesicles isolated from rabbit kidney proximal tubule. Biochim Biophys Acta 1989; 986: 332-340.

231. Van der Goot F, Ripoche P, Corman B. Determination of solute reflection coefficients in kidney brush-border membrane vesicles by light scattering: influence of refractive index. Biochim Biophys Acta 1989; 979: 272-274.

232. Van der Goot F, Seigneur A, Gaucher JC, Ripoche P. Flow cytometry and sorting of amphibian bladder endocytic vesicles containing ADH-sensitive water channels. J Membr Biol 1992; 128: 133-139.

233. Van Heeswijk MPE, Van Os CH. Osmotic water permeabilities of brush border and basolateral membrane vesicles from rat renal cortex and small intestine. J Membr Biol 92: 183-193.

234. Van Hoek AN, de Jong MD, Van Os CH. Effects of dimethylsulfoxide and mercurial sulfhydryl reagents on water and solute permeability of rat kidney brush-border membranes. Biochim Biophys Acta 1990; 1030: 203-210.

235. Van Hoek AN, Hom ML, Luthjens LH, de Jong MD, Dempster JA, Van Os CH. Functional unit of 30 kDa for proximal tubule water channels as revealed by radiation inactivation. J Biol Chem 1991; 226: 16633-16635.

236. Van Hoek AN, Luthjens LH, Hom ML, Van Os CH, Dempster JA. A 30 kDa functional size for the erythrocyte water channel determined in situ by radiation inactivation. Biochem Biophys Res Comm 1992; 184: 1331-1335.

237. Van Hoek AN and Verkman AS. Functional reconstitution of the isolated erythrocyte water channel CHIP28. J Biol Chem 1992; 267: 18267-18269.

238. Van Hoek AN, Wiener M, Bicknese S, Miercke L, Biwersi J, Verkman AS. Secondary structure analysis of purified CHIP28 water channels by CD and FTIR spectroscopy. Biochem 1993; in press.

239. Van Hoek AN, Wiener MC, Lipniunas P, Townsend RR, Farinas J, Verkman AS. Biochemical analysis of CHIP28 structure. JASN 1993; 4: 861 (abstr).

240. Verbavatz JM, Brown D, Sabolic I, Valenti G, Van Hoek AN, Ma T, Verkman AS. Tetrameric assembly of CHIP28 water channels in liposomes and cell membranes. A freeze-fracture study. J Cell Biol 1993; in press.

241. Verbavatz JM, Calamita G, Hugon JS, Bourguet J. Isolation of large sheets of apical material from frog urinary bladder epithelial cells by freeze-fracture. Biol Cell 1989; 66: 91-97.

242. Verbavatz JM, Frigeri A, Gobin R, Ripoche P, Bourguet J. Effects of salt acclimation on water and urea permeabilities across the frog bladder-relationship with intramembrane particle aggregates. Comp Biochem Physiol 1992; 101: 827-833.

243. Verbavatz JM, Thoumine O, Bourguet J. Identification of apical proteins in-

volved in the antidiuretic response of the frog urinary bladder. In: Vasopressin Jard S, Jamison R, eds. Colleque INSERM 1991; 208: 105-115.

244. Verbavatz JM, Van Hoek AN, Ma T, Sabolic I, Valenti G, Ausiello DA, Verkman AS, Brown D. Orthogonal arrays of intramembrane particles in collecting duct principle cells. Submitted.

245. Verkman, AS. Development and biological applications of chloride-sensitive fluorescent indicators. Am J Physiol 1990; 259: C375-C388.

246. Verkman AS. Mechanisms and regulation of water permeability in renal epithelia. Am J Physiol 1989; 257: C837-C850.

247. Verkman, AS. New microfluorimetry approaches to examine cell dynamics. Comments Mol Cell Biophys 1991; 7: 173-187.

250. Verkman AS. Passive H^+/OH^- permeability in epithelial brush border membranes. J Bioenerg Biomembr 1987; 19: 481-493.

251. Verkman AS. Water channels in cell membranes. Annu Rev Physiol 1992; 54: 97-108.

252. Verkman AS, Armijo M, Fushimi K. Construction and evaluation of a frequency-domain epifluorescence microscope for lifetime and anisotropy decay measurements in subcellular domains. Biophys Chem 1991; 40: 117-125.

253. Verkman AS, Dix JA, Seifter JL. Water and urea transport in renal microvillus membrane vesicles. Am J Physiol 1985; 248: F650-F655.

254. Verkman AS, Dix JA, Seifter JL, Skorecki KL, Jung CY, Ausiello DA. Radiation inactivation studies of renal brush border water and urea transport. Am J Physiol 1985; 249: F806-F812.

255. Verkman AS, Fushimi K, Kuwahara M, Shi L-B. New fluorescence methods to examine the regulation of water permeability of kidney collecting tubule and toad urinary bladder by vasopressin. In: The Frontiers of Nephrology. Berliner RW, Honda N, Ullrich KJ, eds. Elsevier Sci Pub 1990; 195-204.

256. Verkman AS, Ives HE. Water transport and fluidity in renal basolateral membranes. Am J Physiol 1986; 250: F633-F643.

257. Verkman AS, Lencer W, Brown D, Ausiello DA. Endosomes from kidney collecting tubule contain the vasopressin-sensitive water channel. Nature 1988; 333: 268-269.

258. Verkman AS, Ma T, Van Hoek AN, Hasegawa H. Structure and function of kidney water channels. In: Gross P, Robertson GL, eds. Vasopressin. 1993; in press.

259. Verkman AS, Masur SK. Very low water permeability and fluidity in granules isolated from toad bladder. J Membr Biol 1988; 104: 241-251.

260. Verkman AS, Skorecki KL, Ausiello DA. Radiation inactivation of multimeric enzymes: application to subunit interactions of adenylate cyclase. Am J Physiol 1986; 250: C103-C115.

261. Verkman AS, Skorecki K, Ausiello DA. Radiation inactivation of oligomeric enzyme systems: theoretical considerations. Proc Natl Acad Sci USA 1984; 81: 150-154.

262. Verkman AS, Van Hoek AN, Zhang R. Identification and molecular cloning of water transporting protiens. In: Alfred Benzon Symp 34, Ussing HH, Fischbarg J, Sten-Knudsen O, Larsen E-H, Willumson N-J, Thaysen JH, eds. Isotonic Transport in Leaky Epithelia. Munksgaard Intl Pub 1993: 388-395.

263. Verkman AS, Weyer P, Brown D, Ausiello DA. Functional water channels are present in clathrin coated vesicles from bovine kidney but not from brain. J Biol Chem 1989; 264: 20608-20613.

264. Verkman AS, Wong K. Proton NMR measurement of diffusional water permeability in suspended renal proximal tubule. Biophys J 1987; 51: 717-723.
265. Verkman AS, Zhang R, Wang Y-X, Shi L-B. The vasopressin sensitive water channel in toad bladder: functional localization in endosomes and mRNA expression in Xenopus oocytes. In: Vasopressin, Jard S and Jamison R, eds. John Libbey Eurotext Ltd 1991; 208: 85-93.
266. Wade JB, Guckian V, Koeppen I. Development of antibodies to apical membrane constituents associated with the action of vasopressin. Curr Top Membr Transp 1984; 20: 217-234.
267. Wade JB, McCusker C, Coleman RA. Evaluation of granule exocytosis in toad urinary bladder. Am J Physiol 1986; 251: C380-C386.
268. Wade JB, Stetson DL, Lewis JA. ADH-action: evidence for a membrane shuttle mechanism. Ann NY Acad Sci 1981; 372: 106-117.
269. Wall SM, Han JS, Chou CL, Knepper MA. Kinetics of urea and water permeability activation by vasopressin in rat terminal IMCD. Am J Physiol 1992; 262: F989-98.
270. Walter P, Blobel G. Preparation of microsomal membranes for cotranslational protein purification. In: Methods of Enzymolmology. New York: Academy Press,1983: 84-93.
271. Wang Y-X, Shi L-B, Verkman AS. Functional water channels and proton pumps are in separate populations of endocytic vesicles from toad bladder granular cells. Biochem 1991; 30: 2888-2894.
272. Welling DJ, Welling LW. Model of renal cell volume regulation without active transport: the role of the heteroporous membrane. Am J Physiol 1988; 255: F529-38.
273. Welling LW, Welling DJ, Ochs TJ. Video measurement of basolateral NaCl reflection coefficient in proximal tubule. Am J Physiol 1987; 253: F290-F298.
274. Welsh MJ, Liedtke CM. Chloride and potassium channels in cystic fibrosis airway epithelia. Nature 1986; 322: 467-470.
275. Welsh MJ, Smith PL, Fromm M, Frizzell RA. Crypts are the site of intestinal fluid and electrolyte secretion. Science 1982; 218: 1219-1221.
276. Wessels HP, Spiess M. Insertion of a multispanning membrane protein occurs sequentially and requires only one signal sequence. Cell 1988; 55: 61-70.
277. Whittembury G, Capri-Medina P, Gonzales E, Linares H. Effect of para-chloromercuribenzenesulfonic acid and temperature on cell water osmotic permeability of proximal straight tubules. Biochim Biophys Acta 1984; 775: 365-373.
278. Williams JC Jr, Abrahamson DR, Schafer JA. Structural changes induced by osmotic water flow in rabbit proximal tubule. Kid Internatl 1991; 39: 672-83.
279. Wistow GJ, Pisano MM, Chepelinski AB. Tandem sequence repeats in transmembrane channel proteins. TIBS 1991; 16: 170-171.
280. Wong KR and Verkman AS. Nuclear magnetic resonance measurement of diffusional water permeability in human platelets. Am J Physiol 1987; 252: C618-C622.
281. Worman HJ, Brasitus TA, Dudeja PK, Fozzard HA, Field M. Relationship between lipid fluidity and water permeability of bovine tracheal epithelial cell apical membranes. Biochem 1986; 25: 1549-1555.
282. Worman HJ, Field M. Osmotic water permeability of small intestinal brush-border membranes. J Membr Biol 1985; 87: 233-239.

283. Ye R, Shi LB, Lencer W, Verkman AS. Functional colocalization of water channels and proton pumps on endocytic vesicles from proximal tubule. J Gen Physiol 1989; 93: 885-902.
284. Ye R, Verkman AS. Osmotic and diffusional water permeability measured simultaneously in cells and liposomes. Biochem 1989; 28: 824-829.
285. Yost CS, Hedgpeth J, Lingappa VR. A stop transfer sequence confers predictable transmembrane orientation to a previously secreted protein in cell-free systems. Cell 1983; 34: 759-766.
286. Zeidel ML, Albalak A, Grossman E, Carruthers A. Role of glucose carrier in human erythrocyte water permeability. Biochem 1992; 31: 589-96.
287. Zeidel ML, Ambudkar SV, Smith BL, Agre P. Reconstitution of functional water channels in liposomes containing purified red cell CHIP28 protein. Biochem 1992; 31: 7436-7440.
288. Zeidel ML, Hammond T, Botelho B, Harris HW. Functional and structural characterization of endosomes from toad bladder epithelial cells. Am J Physiol 1992; 263: F62-F76.
289. Zeidel ML, Strange K, Emma F, Harris HW. Mechanisms and regulation of water transport in the kidney. Sem Neph 1993; 13: 155-167.
290. Zen K, Biwersi J, Periasamy N, Verkman AS. Second messengers regulate endosomal acidification in Swiss 3T3 fibroblasts. J Cell Biol 1992; 119: 99-110.
291. Zeuthen T. Secondary active transport of water across ventricular cell membrane of choroid plexus epithelium of Necturus maculosus. J Physiol 1991; 444: 153-173.
292. Zhang R, Alper S, Thorens B, Verkman AS. Evidence from oocyte expression that the erythrocyte water channel is distinct from band 3 and the glucose transporter. J Clin Invest 1991; 88: 1553-1558.
293. Zhang R, Fushimi K, Lei D, Verkman AS. Expression cloning of water transporting proteins from rat kidney papilla. JASN 1991; 2:730.
294. Zhang R, Logee K, Verkman AS. Expression of mRNA coding for kidney and red cell water channels in *Xenopus* oocytes. J Biol Chem 1990; 265: 15375-15378.
295. Zhang R, Skach W, Hasegawa H, Van Hoek AN, Verkman AS. Cloning, functional analysis and cell localization of a kidney proximal tubule water transporter homologous to CHIP28. J Cell Biol 1993; 120: 359-369.
296. Zhang R, Van Hoek AN, Biwersi J, Verkman AS. A point mutation at cysteine 189 blocks the water permeability of rat kidney water channel CHIP28k. Biochem 1993; 32: 2938-2941.
297. Zhang R and Verkman AS. Urea transport in freshly isolated and cultured cells from rat inner medullary collecting duct. J Membr Biol 1990 115; 253-261.
298. Zhang R and Verkman AS. Water and urea transport in Xenopus oocytes: expression of mRNA from toad urinary bladder. Am J Physiol 1991; 260: C26-C34.

INDEX

Items in italics denote figures (f) or tables (t).

A

Agre P, 47
Alveolar epithelium, 92, 96
Aminonaphthalene trisulfonic acid (ANTS), 14, 17
Amphipathic periodicity
  CHIP28, *61f*
  MIP26 family, *61f, 77f*
Amphotericin B water channels, 9, 18, 26, 53
Anion exchanger (AE1). See Anion transporters.
Anion transporters, 45, *46f*
Anisotropy imaging, 29-32
Antidiuretic hormone (ADH), 1. See also Vasopressin.
ATPase, 2

B

Basolateral integral protein (BLIP), 82
Benga G, 40
Bentley PJ, 1
Big brain protein (BiB), 74, 78
Bourguet J, 41
Brown PA, 39

C

Chang CT, 62
Channel-forming integral protein (CHIP28), 3, 33, 34, 47, 48, 51-53, *54f*
  collecting duct water channel, 80-81
  functional unit, 68-69
  glycosylation, 58-59
  hydropathy analysis, *76f*
  inhibition by mercurials, 56-57
  mammalian cells, 55-56
  nonwater transporting roles, 97
  purification and reconstitution, 48, *49f,* 50-51
  secondary structure analysis
    hydropathy, 60-61
    spectroscopy, 62-64
      tryptophan environment and fluorescence, 64-65
  tetramers in membranes, 65-68
  tissue distribution
    mammalian kidney, 85-88, *89f,* 90
    nonrenal, 90
      airway and alveoli, 92, 96
      choroid plexus, 91-92, *93f*
      ciliary body, 92, 94
      colon, 94, *95f*
      cornea, 92
      iris, 94
      male reproductive tract, 95-96
      pancreas, 94
      skin, 94-95
      spleen, 96
  transmembrane topology, 69-73
  Xenopus oocytes, 53, 55

Index

Choroid plexus, 91-92, *93f*
Chou-Fasman turn propensities, 61, 76
Ciliary body, 92, 94
Circular dichroism (CD), 62, *63f*
Clathrin-coated pits, 21, 38
Colon, 94, *95f*
Cornea, 92
Cystic fibrosis transmembrane regulator (CFTR), 45, 47, 92
Cytoplasmic viscosity, *32f*

D
DER2, 97
Diffusional water permeability coefficient (P_d), 7, 13-14, 16-17
Donnelly D, 76

E
Endosomes, 13
    proximal tubules, 36-38
    vasopressin
        kidney tubules, 19-22
        membrane shuttle hypothesis, 1-2, 19-21
        toad bladder, 23-26
        trafficking mechanism, 26-27
Erythrocytes, 4
    anion transporter, 45, *46f*
    water transport, 33-34, 97

F
Fischbarg J, 17, 45, 46
Fourier transform infrared (FTIR), 62-64
Frigeri A, 41, 81
Fushimi K, 31, 80

G
Gamicidin A water channels, 9, 26
Glucose transporter 1 (GLUT1), 17, 46-47
Glycerol facilitator protein (GlpF), 74, 77, 78
Grantham JJ, 29

H
Harris HW, 24, 26, 40, 82
Harvey B, 26
Hasegawa H, 82
Hess P, 69
House CR, 2

I
Intramembrane particles (IMPs), 4, 65-68
Iris, 94

J
Jähnig F, 60

K
Kedem-Katchalsky nonequilibrium thermodynamic equations, 7, 10, 39
Kidney
    collecting duct
        principal cells, 82-83
    loop of Henle, 85-86
    proximal tubule, 34-3
        solute reflection coefficient, 38-39
        subcellular vesicles, 36-38
Kyte-Doolittle hydropathy value, 60, 76

L
Lanahan A, 97
LAW2, 96, 97
Leaf A, 1
Le Grimellec C, 29
Lysosomes, 27, *28f*

M

Macey RI, 33, 34
Major intrinsic protein 26 (MIP26), 52, 61
  family, 74-78
    mammalian tissues, 96-97
    PCR cloning of homologous proteins, 78-82
    water transport properties, 78
Male reproductive tract, 95-96
Masters BR, 29
Maurel C, 78
Membrane shuttle hypothesis, 1-2, 19-21
Mercurial sulfhydryl compounds, 9
Mitra A, 73
Muallem S, 14

N

Nephron
  water transport, *3f*
NOD26 (soybean nodulin), 74, 78

O

Onsager reciprocity relation, 6
Orthogonal arrays of particles (OAP), 83
Osmotic water permeability
  measurement in cells and vesicles
    flourescence quenching, 11, *12f*, 13
    kidney tubules, 15-17
    light scattering, 9-11
    single cell, 14-15
    Xenopus oocytes, 17-18

P

Pancreas, 94
Park K, 62
Permeability coefficient (P_f), 5-6, 11, 15
Pfeffer W, 1
Pietras RJ, 29
Proton pumps, 26, 27, 37-38

S

Shi L-B, 26, 69
Skach W, 69
Skin, 94-95
Solomon AK, 1, 34, 40
Solute reflection coefficient, 6
  proximal tubule, 38-39
Soybean nodulin. See NOD26.
Spleen, 96
Stern-Volmer constants, 65
Stroud R, 73

T

Tobacco mosaic protein (ToB), 74, 78
Tonoplast intrinsic protein (TIP), 74, 78
Tracheobronchial epithelium, 92
Turgor response protein (TUR), 74, 78

U

Urea, 25-26, 34, 97

V

Valenti G, 40
Van der Groot F, 40
Van Hoek AN, 41, 73
Van Os CH, 41
Vasopressin
  amphibian urinary bladder, *4f*
  nephron, 3-4, 16
  regulation of transepithelial water permeability, 19-21
  vesicular trafficking mechanism, 26-27
    regulation by osmotic gradients, 27-29
  water permeability and cell fluidity, 29-32
  water permeabilty in endosomes from kidney collecting duct, 21-22, *23f*
  water permeabilty in endosomes from toad bladder, 21-22, *23f*
  water permeabilty in toad bladder granules, 26
Verbavatz JM, 40
Verkman AS, 69
Verma A, 78

W

Water transport
  biophysics, 5-9
  extrarenal membranes, 39
  inhibition by mercurials, 56-57
  measurement
    diffusional permeability, 13-14
    osmotic permeability
      flourescence quenching,
        11, *12f*, 13
      kidney tubules, 15-17
      light scattering, 9-11
      single cells, 14-15
    Xenopus oocytes, 17-18
  membrane characteristics and, 5t
  major intrinsic protein 26 (MIP26),
    78
  putative transporters
    anion transporters and, 45
    antibody labeling, 41
    biochemical labeling, 39-40
    determination of size by target
      analysis, 41-42
    glucose transporters and, 46-47

WCH3, 97
WCH-CD, 80-81
  mammalian kidney, 85, *89f,* 90, 97
Whittembury M, 35
Wiener M, 73
Wright EM, 29

X

Xenopus oocytes
  water channels
    channel-forming integral protein
      (CHIP28), 53, 55
      functional unit, 68-69
    expression, 42-45
    measurement, 17-18

Y

Yaeger M, 73

Z

Zeidel ML, 24